Vertical Farming in Urban Areas: Innovative Agriculture, Food Security, and Sustainable Growth Practices

Vertical Farming in Urban Areas: Innovative Agriculture, Food Security, and Sustainable Growth Practices

Introduction

Chapter 1: The Basics of Vertical Farming

I0477867

Chapter 2: Design and Infrastructure

Chapter 3: Economic Aspects of Vertical Farming

Chapter 4: Environmental Impact

Chapter 5: Social and Cultural Implications

Chapter 6: Challenges and Solutions

Chapter 7: Future of Vertical Farming

Conclusion

Introduction

Vertical farming is revolutionizing the way we think about agriculture, bringing sustainable and innovative food production directly into urban environments. As cities continue to grow and the demand for fresh, locally sourced produce increases, vertical farming offers a solution that addresses food security and promotes environmental sustainability and community well-being. By utilizing advanced technologies and optimizing resource use, vertical farms can produce high yields in small spaces, reducing the environmental impact of traditional agriculture. This book explores the multifaceted benefits and challenges of vertical farming, highlighting its potential to transform urban food systems and contribute to a more resilient and sustainable future.

Overview of Vertical Farming

Vertical farming represents a groundbreaking approach to agriculture, offering a sustainable and efficient solution to the challenges of urban food production.

Definition and Concept

Vertical farming is an innovative agricultural practice that involves growing crops in vertically stacked layers or integrated structures, such as skyscrapers, warehouses, or shipping containers. This approach leverages controlled-environment agriculture (CEA) technology to optimize plant growth by precisely managing light, temperature, humidity, and nutrient delivery variables. The primary objective of vertical farming is to maximize food production in urban areas while minimizing the use of land, water, and other natural resources.

The concept of vertical farming involves producing food in proximity to urban centers, thereby reducing the need for extensive transportation networks and decreasing the associated carbon footprint. This method also allows year-round cultivation, irrespective of external weather conditions, ensuring a consistent and reliable food supply.

Vertical farming systems can be categorized into several types based on the techniques used. Hydroponics involves growing plants in a nutrient-rich water solution, eliminating the need for soil. On the other hand, Aeroponics grows plants in an air or mist environment, delivering nutrients directly to the plant roots through a fine mist. Aquaponics combines hydroponics with aquaculture, where waste produced by farmed fish supplies nutrients for the plants, and the plants, in turn, help purify the water for the fish.

Another critical component of vertical farming is using artificial lighting, typically LED lights. These lights provide the necessary spectrum for photosynthesis, allowing plants to grow in environments with little to no natural sunlight. Vertical farms often employ advanced technologies such as the Internet of Things (IoT), artificial intelligence (AI), and robotics to monitor and manage plant health, optimize resource use, and automate harvesting processes.

In summary, vertical farming represents a paradigm shift in agricultural practices, offering a sustainable and efficient solution to meet the growing food demands of urban populations. By integrating advanced technologies and innovative design, vertical farming can significantly enhance food security, reduce environmental impact, and contribute to developing resilient urban food systems.

Historical Background

Vertical farming is not entirely new; its roots can be traced back to ancient civilizations. The Hanging Gardens of Babylon, one of the Seven Wonders of the Ancient World, are often cited as an early example of vertically arranged agriculture. These terraced gardens were said to have been constructed around 600 BCE in what is now Iraq, showcasing an early form of stacked cultivation.

In the modern era, vertical farming began in the early 20th century. American geologist Gilbert Ellis Bailey coined "vertical farming" in his 1915 book "Vertical Farming," where he envisioned using tall buildings to grow food. However, it wasn't until the late 20th and early 21st centuries that technological advancements made vertical farming a feasible and scalable solution.

Dr. Dickson Despommier, a professor at Columbia University, played a pivotal role in popularizing the concept in the early 2000s. His vision of vertical farming involved transforming urban skyscrapers into multi-story farms that could produce significant quantities of food for city dwellers. Despommier's work brought considerable attention to the potential benefits of vertical farming, including reducing the pressure on rural land, conserving water, and cutting down on transportation emissions.

The first commercial vertical farms began to emerge in the early 2010s. Pioneering companies like AeroFarms in the United States and Mirai in Japan developed large-scale vertical farming operations, utilizing state-of-the-art technology to grow leafy greens and herbs in controlled indoor environments. These early ventures demonstrated the viability of vertical farming as a commercial enterprise and laid the groundwork for further innovations in the field.

Today, vertical farming continues to evolve, driven by technological advancements and growing interest in sustainable urban agriculture. Its historical journey, from ancient innovations to cutting-edge modern practices, highlights its potential to revolutionize food production in urban environments.

Current Trends and Statistics

Vertical farming is experiencing rapid growth and innovation, driven by the need for sustainable food production solutions in urban areas. As of the early 2020s, the global vertical farming market is valued at several billion dollars and is projected to expand significantly in the coming years. This growth is fueled by increasing urbanization, rising awareness of environmental issues, and advancements in agricultural technology.

One of the notable trends in vertical farming is the diversification of crops. While early vertical farms primarily focused on leafy greens and herbs due to their fast growth cycles and high market demand, there is now a growing interest in cultivating a broader range of crops, including fruits, vegetables, and even staple grains. This diversification aims to enhance food security and provide urban populations with a wider variety of fresh, locally grown produce.

Technological innovation continues to play a critical role in the development of vertical farming. Advanced LED lighting systems, automated nutrient delivery, and sophisticated climate control technologies are becoming standard features in modern vertical farms. Additionally, integrating IoT and AI allows for precise monitoring and optimization of growing conditions, resulting in higher yields and reduced resource consumption.

Statistics highlight the impressive productivity of vertical farming compared to traditional agriculture. For example, vertical farms can produce up to 10 times more food per square meter than conventional farms, thanks to their multi-layered growing structures and efficient use of resources. Water usage in vertical farming is also significantly lower, with some systems using up to 95% less water than traditional soil-based agriculture.

Moreover, vertical farming is gaining traction in various regions around the world. Countries like Singapore, Japan, and the United Arab Emirates invest heavily in vertical farming initiatives to enhance food security and reduce dependence on imported produce. As urban populations continue to grow, the adoption of vertical farming is expected to increase, positioning it as a key component of sustainable urban development.

In conclusion, the current trends and statistics in vertical farming underscore its potential to address the challenges of urban food production. By leveraging cutting-edge technology and innovative agricultural practices, vertical farming offers a viable solution for sustainably feeding the growing urban population.

Importance of Vertical Farming

Vertical farming is important because it has the potential to revolutionize food production, enhance urban sustainability, and address critical challenges related to food security, land use, and environmental impact.

Food Security

Food security is a critical global issue, with the world's population expected to reach nearly 10 billion by 2050. This dramatic increase places immense pressure on traditional agricultural systems to produce enough food to meet the growing demand. Vertical farming offers a promising solution to enhance food security, especially in urban areas where land availability and food distribution can be significant challenges.

One of the primary advantages of vertical farming in addressing food security is its ability to produce high yields in limited spaces. By utilizing vertically stacked layers, vertical farms can grow a substantial amount of food in a fraction of the land required for conventional farming. This efficiency is particularly beneficial in densely populated urban areas, where land is scarce and expensive. Vertical farms can be established in unused urban spaces, such as abandoned buildings, rooftops, and warehouses, transforming these areas into productive agricultural sites.

Vertical farming also contributes to food security by providing a consistent and reliable food supply. Traditional farming is highly dependent on weather conditions, which can be unpredictable and increasingly erratic due to climate change. In contrast, vertical

6

farming operates in controlled environments, where factors such as temperature, humidity, light, and nutrients are precisely managed. This control minimizes the risk of crop failures and ensures year-round production, regardless of external weather conditions.

Moreover, vertical farming can significantly reduce the food supply chain's vulnerability to disruptions. Urban vertical farms are located close to consumers, reducing the need for long transportation routes and the associated risks of spoilage and delays. This proximity ensures fresher produce and strengthens local food systems, making them more resilient to global supply chain disruptions, such as those experienced during the COVID-19 pandemic.

In addition, vertical farming can play a vital role in addressing food deserts—areas with limited access to fresh, nutritious food. By establishing vertical farms in these underserved communities, residents can gain access to locally grown, affordable produce, improving their overall food security and health outcomes. In summary, vertical farming offers a sustainable and efficient approach to enhancing food security, particularly in urban areas, by maximizing land use, ensuring consistent production, and strengthening local food systems.

Urbanization and Land Use

Urbanization is a global trend that continues to accelerate, with more than half of the world's population now living in cities. This rapid urbanization poses significant challenges for land use and food production. As cities expand, agricultural land is often converted into residential, commercial, and industrial areas, reducing available farmland. Vertical farming addresses these challenges by offering a sustainable way to integrate agriculture into urban environments.

One key benefit of vertical farming in urbanization is its ability to use space efficiently. Traditional agriculture requires vast tracts of land to produce sufficient food, but vertical farming maximizes space utilization by growing crops in stacked layers. This vertical approach allows for high-density food production within the urban footprint, reducing the need to expand cities into surrounding rural areas.

Vertical farming also helps to preserve natural ecosystems and biodiversity by minimizing the encroachment of agricultural activities on forests, wetlands, and other critical habitats. By concentrating food production in urban areas, vertical farming reduces the pressure to clear additional agricultural land, contributing to conserving these valuable ecosystems.

Moreover, vertical farms can be integrated into existing urban infrastructure, such as converting unused buildings or retrofitting rooftops and walls. This integration not only makes efficient use of space but also revitalizes underutilized urban areas, creating green spaces that enhance the aesthetic and environmental quality of cities. Vertical farming can also be incorporated into urban planning and development projects, promoting sustainable land use practices.

In conclusion, vertical farming provides an innovative solution to the challenges posed by urbanization and land use. By maximizing space efficiency and integrating agriculture into urban environments, vertical farming helps to sustain food production without compromising natural ecosystems and contributes to the creation of greener, more sustainable cities.

Environmental Benefits

Vertical farming offers a range of environmental benefits that make it a sustainable alternative to traditional agricultural practices. One of the most significant advantages is the substantial reduction in water usage. Traditional agriculture is notorious for its high water consumption, with irrigation accounting for about 70% of global freshwater withdrawals. In contrast, vertical farming uses advanced hydroponic, aeroponic, and aquaponic systems that require up to 95% less water. These systems recycle water and nutrients, ensuring efficient use and minimal waste, which is crucial in the face of growing water scarcity.

Another key environmental benefit of vertical farming is the reduction of pesticide and herbicide use. Because vertical farms operate in controlled environments, they can effectively manage pests and diseases without relying on chemical treatments. This approach not only produces healthier food but also prevents the contamination of soil and water sources, which is a common issue with conventional farming.

Vertical farming also contributes to lower greenhouse gas emissions. Traditional agriculture involves significant energy consumption and emissions from activities such as plowing, planting, and harvesting, as well as the transportation of produce over long distances. Vertical farms, on the other hand, often utilize energy-efficient LED lighting and can be powered by renewable energy sources, such as solar panels. Additionally, their urban locations reduce the need for extensive transportation, decreasing the carbon footprint associated with food production and distribution.

Furthermore, vertical farming can help mitigate the urban heat island effect. Vertical farms can help cool cities and improve air quality by integrating green spaces and vegetation into urban areas. Plants naturally absorb carbon dioxide and release oxygen, creating a healthier urban environment.

In summary, vertical farming presents a range of environmental benefits, including reduced water usage, decreased reliance on chemicals, lower greenhouse gas emissions, and improved urban climate resilience. These advantages make vertical farming a compelling and sustainable option for future food production.

Chapter 1: The Basics of Vertical Farming

Vertical farming is an innovative approach to agriculture that reimagines how we grow food in urban environments. By utilizing vertical space and advanced technology, it offers a sustainable solution to the challenges of traditional farming, such as land scarcity and environmental degradation. This chapter will explore the fundamental aspects of vertical farming, including its definition, various systems, and key components. We will delve into the technological foundations enabling vertical farming, from hydroponics to IoT and AI, providing a comprehensive understanding of this transformative agricultural practice.

What is Vertical Farming?

Vertical farming is a revolutionary method of agricultural production that aims to address the challenges of traditional farming by growing crops in vertically stacked layers within controlled environments. This innovative approach leverages advanced technologies to optimize resource use, increase productivity, and produce fresh, nutritious food closer to urban populations. Vertical farming offers a sustainable solution to the growing demands for food in urban areas while reducing the environmental impact associated with conventional farming practices.

Types of Vertical Farming Systems

Vertical farming systems can be categorized based on the techniques used to grow crops. The three primary types of vertical farming systems are hydroponics, aeroponics, and aquaponics, each with distinct methods for delivering nutrients and supporting plant growth.

Hydroponics

In hydroponic systems, plants are grown in a nutrient-rich water solution rather than soil. This method allows for precise control over nutrient delivery, ensuring that plants receive the nutrients they need for optimal growth. Hydroponic systems can be further divided into several subtypes, including nutrient film technique (NFT), deep water culture (DWC), and drip systems. NFT involves a thin film of nutrient solution flowing over the roots, while DWC suspends plant roots in oxygenated water. Drip systems deliver nutrient solution directly to the base of each plant. Hydroponics is widely used in vertical farming due to its efficiency, reduced water usage, and high yields.

Aeroponics

Aeroponic systems grow plants in an air or mist environment, where the roots are suspended in the air and periodically misted with a nutrient-rich solution. This method offers several advantages, including increased oxygen availability to the roots, faster growth rates, and reduced water and nutrient usage. Aeroponics is particularly effective for growing plants with extensive root systems and is often used in vertical farms to maximize space and resource efficiency.

Aquaponics

Aquaponics combines hydroponics with aquaculture, creating a symbiotic relationship between plants and fish. In this system, fish waste provides essential nutrients for the plants, while the plants help filter and purify the water for the fish. Aquaponics is a closed-loop system miming natural ecosystems, promoting sustainability and resource conservation. This method is beneficial for producing crops and fish, offering a diverse range of products from a single system.

These vertical farming systems offer unique benefits and can be tailored to specific crop types, environmental conditions, and production goals. The system choice depends on available space, budget, and desired crop varieties, making vertical farming a versatile and adaptable approach to urban agriculture.

Key Components

Vertical farming relies on several key components to create a controlled environment that optimizes plant growth and resource use. These components work together to provide the ideal conditions for cultivating crops in vertical layers:

- Growing Structures: The physical framework of a vertical farm consists of shelves, racks, or towers that hold the plants in vertically stacked layers. These structures can be customized to fit various spaces, including repurposed buildings, shipping containers, and greenhouses. The design of the growing structures maximizes space utilization and allows for easy access to the plants for maintenance and harvesting.
- Lighting Systems: Artificial lighting is essential in vertical farming to provide the necessary light spectrum for photosynthesis. LED lights are commonly used due to their energy efficiency, longevity, and ability to emit specific wavelengths of light that promote plant growth. LED lighting systems can be adjusted to mimic natural sunlight, providing the optimal light intensity and duration for different stages of plant development.
- Climate Control: Maintaining a stable environment is crucial for the success of vertical farming. Climate control systems regulate the growing area's temperature, humidity, and ventilation. Advanced sensors and automated systems monitor and adjust these parameters in real-time, ensuring that plants remain in their ideal growth conditions. Proper climate control helps prevent diseases, pests, and other stress factors affecting crop yield and quality.
- Nutrient Delivery Systems: Depending on the vertical farming system, various methods deliver nutrients to the plants. In hydroponics, a nutrient solution is circulated through the system to provide essential minerals and nutrients to the roots directly. Aeroponic systems use misting devices to spray the nutrient solution onto the roots, while aquaponics relies on fish waste as a natural nutrient source.

These systems ensure that plants receive a balanced and consistent supply of nutrients for optimal growth.

- Water Management: Efficient water use is a hallmark of vertical farming. Recirculating water systems reduce waste and conserve resources by reusing water within the system. Advanced filtration and purification technologies ensure that the water remains clean and free of contaminants. Water management systems also include mechanisms for monitoring water quality and levels, preventing shortages and ensuring a continuous plant supply.

- Automation and Monitoring: Automation plays a significant role in vertical farming by streamlining operations and reducing labor costs. Automated systems can handle planting, watering, nutrient delivery, and harvesting tasks. Monitoring systems equipped with IoT sensors collect data on environmental conditions, plant health, and system performance. This data is analyzed in real-time to make adjustments and optimize the growing process, leading to higher yields and better resource efficiency.

By integrating these key components, vertical farming creates a highly controlled and efficient environment for crop production. This approach not only maximizes space and resource use but also ensures consistent and high-quality yields, making it a viable solution for urban agriculture.

Commonly Grown Crops

Vertical farming is particularly well-suited for growing various crops, especially those that benefit from controlled environments and space-efficient cultivation methods. While the choice of crops depends on factors such as market demand, growing conditions, and system capabilities, several types of plants are commonly grown in vertical farms due to their suitability and high yield potential:

- Leafy Greens: Leafy greens such as lettuce, spinach, kale, and arugula are among the most popular crops grown on vertical farms. These plants have relatively short growth cycles, allowing for multiple harvests throughout the year. Leafy greens thrive in hydroponic and aeroponic systems, receiving consistent light and nutrient supply. Their compact size and high market demand make them ideal candidates for vertical farming.

- Herbs: Culinary herbs like basil, mint, parsley, cilantro, and chives are commonly grown in vertical farms. Herbs are well-suited for vertical farming because they require minimal space and can be harvested frequently. The controlled environment of vertical farms allows for year-round production of fresh herbs, which are highly valued for their flavor and nutritional benefits. Growing herbs vertically also reduces the risk of contamination and ensures a steady supply for local markets.
- Microgreens: Microgreens are young, edible plants harvested at an early growth stage, usually within 7-14 days after germination. These nutrient-dense crops include radish, broccoli, sunflower, and mustard greens. Microgreens are ideal for vertical farming due to their rapid growth, high yield per unit area, and strong market demand. They are often grown in stacked trays or shelves, making efficient use of vertical space.
- Fruits and Vegetables: Certain fruits and vegetables can be successfully grown in vertical farms. Tomatoes, strawberries, cucumbers, and peppers are popular choices. These crops may require more space and support structures, but advanced vertical farming systems can accommodate their growth requirements. By providing optimal light, temperature, and nutrient conditions, vertical farms can produce high-quality fruits and vegetables with minimal resource use.
- Specialty Crops: Vertical farming offers the flexibility to grow specialty crops that cater to niche markets. These include medicinal plants, edible flowers, and exotic varieties not commonly available in traditional markets. The controlled environment of vertical farms allows for precise cultivation of these specialty crops, ensuring consistent quality and supply.

In conclusion, vertical farming is a versatile and efficient method for growing various crops. From leafy greens and herbs to fruits and specialty plants, vertical farms can produce high-quality, nutritious food year-round in urban environments. By optimizing space and resources, vertical farming contributes to food security and sustainability, making it a promising solution for the future of agriculture.

Technological Foundations

Vertical farming relies heavily on advanced technologies to create efficient and sustainable agricultural systems. These technologies optimize plant growth, maximize resource use, and ensure high-quality yields. This section explores the three main types of vertical farming systems—hydroponics, aeroponics, and aquaponics—as well as the principles of controlled environment agriculture (CEA) and the role of the Internet of Things (IoT) and artificial intelligence (AI) in modern vertical farms.

Hydroponics, Aeroponics, and Aquaponics

The success of vertical farming hinges on the innovative use of soilless cultivation methods. Hydroponics, aeroponics, and aquaponics are the primary techniques, each offering unique benefits and applications.

Hydroponics

Hydroponics is a method of growing plants in a nutrient-rich water solution, eliminating the need for soil. This system directly provides plants with essential nutrients through the water, allowing for precise control over nutrient delivery and promoting faster growth rates. There are several subtypes of hydroponic systems, including:

- Nutrient Film Technique (NFT): Plants are grown in shallow channels where a thin film of nutrient solution continuously flows over the roots. This method ensures that roots receive adequate oxygen and nutrients.

- Deep Water Culture (DWC): Plants are suspended in net pots above a reservoir of oxygenated nutrient solution, with their roots submerged in the water. This system is particularly effective for leafy greens and herbs.
- Drip System: Nutrient solution is delivered directly to the base of each plant through a network of tubes and emitters. Excess solution is collected and recirculated, minimizing waste.

Hydroponic systems are highly efficient, using up to 90% less water than traditional soil-based agriculture. They are also scalable and can be adapted to various indoor environments, making them ideal for urban vertical farms.

Aeroponics

Aeroponics involves growing plants in an air or mist environment, where roots are suspended and periodically misted with a nutrient-rich solution. This technique offers several advantages:

- Increased Oxygen Availability: Roots receive ample oxygen, promoting faster and healthier growth.
- Efficient Nutrient Use: Nutrients are delivered directly to the roots in a fine mist, ensuring optimal absorption and reducing waste.
- Space Efficiency: Aeroponic systems require less space than traditional methods, allowing for higher density planting.

Aeroponics is particularly suited for high-value crops and plants with extensive root systems. Its efficiency and productivity are key components of advanced vertical farming operations.

Aquaponics

Aquaponics combines hydroponics with aquaculture, creating a symbiotic relationship between plants and fish. In this system:

- Fish Waste: Fish produce waste that is rich in nutrients. This waste is converted by beneficial bacteria into forms that plants can absorb.
- Plant Filtration: Plants absorb these nutrients, effectively filtering and purifying the water for the fish.
- Closed-Loop System: This creates a sustainable, closed-loop ecosystem where fish and plants thrive.

Aquaponics offers the dual benefit of producing crops and fish, making it a versatile and sustainable method for urban farming. It also reduces the need for chemical fertilizers, as the fish waste provides natural nutrients for the plants.

Controlled Environment Agriculture

Controlled Environment Agriculture (CEA) is the cornerstone of vertical farming, enabling the precise management of environmental conditions to optimize plant growth. CEA encompasses various technologies and practices designed to create ideal growing environments regardless of external weather conditions:

- Climate Control:
    - Temperature Regulation: Maintaining optimal temperatures is crucial for plant health and productivity. Climate control systems use heating, ventilation, and air conditioning (HVAC) to ensure consistent temperatures.
    - Humidity Control: Proper humidity levels prevent plant stress and disease. Dehumidifiers and humidifiers maintain the desired humidity range.
    - Ventilation: Adequate airflow is essential to prevent mold and mildew. Ventilation systems ensure fresh air circulation, providing plants with sufficient carbon dioxide for photosynthesis.

- Lighting:
  - Artificial Lighting: LED lights are commonly used in vertical farming to provide the necessary light spectrum for photosynthesis. LED technology allows for the customization of light wavelengths, intensities, and durations to match the specific needs of different crops.
  - Light Cycles: Controlled lighting simulates day and night cycles, promoting natural growth patterns and maximizing photosynthetic efficiency.
- Nutrient Delivery:
  - Automated Systems: Nutrient delivery systems are automated to ensure precise and consistent plant supply. Sensors monitor nutrient levels, and automated dispensers adjust the concentration and distribution of nutrients based on real-time data.
  - Recirculation: Nutrient solutions are recirculated within the system, minimizing waste and ensuring efficient use of resources.
- Water Management:
  - Water Recycling: Water used in vertical farming systems is recycled and purified, significantly reducing water consumption compared to traditional farming.
  - Irrigation Control: Automated irrigation systems deliver water directly to the plant roots, preventing overwatering and ensuring optimal hydration.

- Pest and Disease Management:
  - Integrated Pest Management (IPM): IPM strategies control pests and diseases without chemical pesticides. This includes biological controls, such as introducing beneficial insects, and physical barriers to prevent pest access.
  - Sanitation Protocols: Strict sanitation protocols are followed to minimize the risk of disease outbreaks. This includes regular cleaning and sterilization of equipment and growing areas.

CEA allows vertical farms to achieve consistent, high-quality yields by creating optimal conditions for plant growth. This approach reduces the impact of external environmental factors and enhances the sustainability and productivity of urban agriculture.

Role of IoT and AI

Integrating the Internet of Things (IoT) and artificial intelligence (AI) is transforming vertical farming by providing advanced tools for monitoring, automation, and data analysis. These technologies enhance efficiency, reduce labor costs, and optimize resource use:

- IoT in Vertical Farming:
    - Sensor Networks: IoT sensors are deployed throughout vertical farms to monitor key parameters such as temperature, humidity, light intensity, soil moisture, and nutrient levels. These sensors collect real-time data, providing valuable insights into the growing environment.
    - Remote Monitoring: IoT-enabled devices allow farmers to monitor and control their systems remotely. This capability is particularly beneficial for managing large-scale operations and ensuring timely responses to any issues.
    - Data Collection and Analysis: The data collected by IoT sensors is analyzed to identify patterns and trends. This information helps farmers make informed decisions about irrigation schedules, nutrient delivery, and environmental adjustments.
- AI in Vertical Farming:
    - Predictive Analytics: AI algorithms analyze historical and real-time data to predict future outcomes. For example, AI can forecast crop yields, detect pest infestations, and predict optimal harvest times.
    - Automation: AI-powered automation systems handle routine tasks such as planting, watering, and harvesting. Robotics and automated machinery reduce

the need for manual labor, increasing efficiency and consistency.
- ○ Optimization Algorithms: AI optimization algorithms continuously adjust environmental conditions to maximize plant growth and resource efficiency. These algorithms consider plant species, growth stages, and environmental variables.
- Smart Farming Platforms:
  - ○ Centralized Management: Smart farming platforms integrate IoT and AI technologies into a centralized system, allowing farmers to manage all aspects of their vertical farm from a single interface. This includes monitoring environmental conditions, controlling automated systems, and analyzing performance data.
  - ○ Decision Support: These platforms provide decision support tools that offer recommendations based on data analysis and predictive models. This helps farmers make informed choices about crop management and resource allocation.

- Sustainability and Efficiency:
  - ○ Resource Optimization: IoT and AI technologies help optimize water, energy, and nutrients, reducing waste and lowering operational costs. For example, precision irrigation systems ensure that plants receive the exact amount of water they need, minimizing excess use.
  - ○ Environmental Impact: IoT and AI contribute to the sustainability of vertical farming by improving resource efficiency and reducing reliance on chemical inputs. These technologies help minimize the environmental footprint of urban agriculture, promoting a more sustainable food production system.

In conclusion, the technological foundations of vertical farming—comprising hydroponics, aeroponics, aquaponics, controlled environment agriculture, and the integration of IoT and AI—are essential for creating efficient, sustainable, and productive urban farming systems. These advanced technologies enable vertical farms to overcome the limitations of traditional agriculture, ensuring consistent, high-quality yields and contributing to the future of sustainable urban food production.

Chapter 2: Design and Infrastructure

The design and infrastructure of vertical farming systems are critical to their success and efficiency. Creating a well-planned vertical farm involves understanding the architectural and structural requirements and integrating advanced technologies and logistical systems that optimize space, energy, and resource use. This chapter will delve into the key architectural considerations, infrastructure components, and logistical frameworks that support the establishment and operation of vertical farms. By exploring these elements, we can better appreciate how innovative design and robust infrastructure contribute to the sustainability and productivity of vertical farming in urban environments.

Architectural Considerations

The architectural design of vertical farms is a fundamental aspect that determines their efficiency, sustainability, and overall success. Proper architectural planning ensures optimal use of space, energy, and resources, making it possible to produce high yields of fresh produce in urban environments. This section covers building structures and retrofitting, lighting and climate control, and space optimization, providing a comprehensive understanding of the architectural considerations involved in vertical farming.

Building Structures and Retrofitting

One of the primary architectural considerations in vertical farming is the choice of building structures. Vertical farms can be established in various buildings, offering unique advantages and challenges. The main options include new constructions, repurposed buildings, and retrofitted existing structures:

- New Constructions:
  - Purpose-Built Facilities: Designing and constructing purpose-built vertical farms allows for customized solutions tailored to the specific needs of vertical

farming. These facilities can incorporate advanced technologies, optimal layouts, and sustainable materials from the outset. Purpose-built structures offer the flexibility to design for maximum efficiency, energy conservation, and scalability.

- Green Building Standards: Incorporating green building standards, such as LEED (Leadership in Energy and Environmental Design), ensures that new vertical farm constructions meet high sustainability and energy efficiency criteria. This approach minimizes environmental impact and promotes long-term operational efficiency.

- Repurposed Buildings:
  - Adaptive Reuse: Repurposing existing buildings, such as warehouses, factories, and office buildings, for vertical farming is a cost-effective and sustainable option. Adaptive reuse leverages existing infrastructure, reducing the need for new construction and minimizing environmental impact. These buildings often have large, open spaces suitable for vertical farming systems.
  - Structural Modifications: While repurposing existing buildings, structural modifications may be necessary to accommodate vertical farming equipment and systems. This can include reinforcing floors to support the weight of growing structures, installing new HVAC systems, and modifying lighting and electrical systems to meet the demands of vertical farming.

- Retrofitting Existing Structures:
  - Rooftop Farms: Retrofitting rooftops of existing buildings is a popular approach for urban vertical farming. Rooftop farms utilize underused space, transforming it into productive agricultural areas. These installations often require structural assessments to ensure the roof can support the additional weight of growing systems and equipment.
  - Vertical Additions: In some cases, vertical additions to existing buildings can create new space for farming

without expanding the building's footprint. This approach maximizes urban space use and integrates farming into the city's architecture.

The choice of building structure and retrofitting options depends on available space, budget, and specific farming needs. Proper architectural planning and structural modifications ensure vertical farms can operate efficiently and sustainably in various urban environments.

Lighting and Climate Control

Lighting and climate control are critical components of vertical farming that directly impact plant growth, health, and productivity. Advanced systems and technologies are employed to create optimal growing conditions, regardless of external weather variations:

- Lighting Systems:
  - LED Lighting: LED (light-emitting diode) lighting is preferred for vertical farming due to its energy efficiency, longevity, and customizable light spectrum. LEDs can be tuned to emit specific wavelengths of light that promote photosynthesis and plant development. This ability to customize light conditions allows for precise control over plant growth cycles, resulting in higher yields and better quality produce.
  - Light Intensity and Duration: Properly managing light intensity and duration is crucial for plant health. Vertical farms use programmable lighting systems to simulate natural daylight cycles, ensuring plants receive the appropriate light for each growth stage. This precise control helps optimize photosynthesis and energy use, leading to faster growth rates and improved crop quality.
- Climate Control Systems:
  - Temperature Regulation: Maintaining a stable temperature is essential for optimal plant growth.

Climate control systems use heating, ventilation, and air conditioning (HVAC) to regulate temperature within the growing environment. Automated sensors and controllers ensure that each crop type's temperature remains within the ideal range.

- o Humidity Control: Proper humidity levels prevent plant stress, reduce disease risk, and promote healthy growth. Dehumidifiers and humidifiers maintain the desired humidity range. Advanced climate control systems monitor and adjust humidity levels in real time, ensuring consistent conditions.
- o Ventilation and Airflow: Adequate ventilation is necessary to provide plants with fresh air and prevent the buildup of heat and humidity. Ventilation systems circulate air within the growing area, ensuring plants receive sufficient carbon dioxide for photosynthesis and reducing the risk of mold and mildew. Proper airflow also helps distribute temperature and humidity evenly throughout the vertical farm.

- Environmental Monitoring:
  - o Sensors and Automation: Environmental sensors continuously monitor key parameters such as light intensity, temperature, humidity, and carbon dioxide levels. This data is analyzed in real-time by automated control systems, which make necessary adjustments to maintain optimal growing conditions. Automation reduces the need for manual intervention, increases efficiency, and ensures consistent crop quality.
  - o Data-Driven Decisions: Advanced climate control systems use data analytics to optimize growing conditions based on historical and real-time data. This approach enables vertical farms to fine-tune their operations, improve resource use, and maximize yields.

In conclusion, effective lighting and climate control are essential for vertical farming's success. Vertical farms can create ideal growing environments that enhance plant health, productivity, and sustainability by leveraging advanced technologies and automation.

Space Optimization

Space optimization is a fundamental aspect of vertical farming, as it efficiently uses limited urban space to maximize food production. Properly designed vertical farms utilize every available inch of space to grow crops, ensuring high yields and resource efficiency:

- Vertical Growing Structures:
  - Stacked Layers: Vertical farms use stacked layers or shelves to grow crops in multiple tiers, significantly increasing the growing area within a confined space. These layers can be arranged vertically in racks or towers, allowing for high-density planting. The design of these structures ensures easy access for planting, maintenance, and harvesting.
  - Modular Systems: Modular growing systems offer flexibility and scalability. These systems consist of individual units that can be easily assembled, disassembled, and reconfigured to accommodate different crops and growing conditions. Modular systems allow vertical farms to expand or adapt their operations as needed, maximizing space use and productivity.
- Innovative Layouts:
  - Rotating Systems: Some vertical farms use rotating growing systems that move plants through different light and climate zones. This approach ensures that all plants receive equal exposure to optimal growing conditions, improving overall yields. Rotating systems also use space efficiently by allowing for continuous production cycles.
  - Vertical Towers: Vertical towers are columnar structures with plants grown in a spiral or stacked configuration around a central support. These towers maximize vertical space and can be integrated into

various urban settings, such as rooftop gardens or indoor farms. They are particularly effective for growing leafy greens and herbs.

- Space-Efficient Crop Selection:
  - Compact Crops: Selecting crops that thrive in compact growing conditions is essential for space optimization. Leafy greens, herbs, and microgreens are well-suited for vertical farming due to their small size and rapid growth cycles. These crops can be densely planted vertically, maximizing yields per unit area.
  - High-Yield Varieties: Vertical farms often cultivate high-yield crop varieties that produce more per square foot than traditional crops. These varieties are bred for their productivity, disease resistance, and suitability for controlled environments.

- Integration with Urban Infrastructure:
  - Rooftop Farms: Utilizing rooftop spaces for vertical farming is an effective way to integrate agriculture into urban environments. Rooftop farms take advantage of unused spaces, transforming them into productive agricultural areas. These farms contribute to urban sustainability by providing fresh produce close to consumers and reducing the heat island effect.
  - Vertical Walls and Facades: Vertical farming can be incorporated into the walls and facades of buildings, creating green walls that produce food while enhancing the aesthetic appeal of urban structures. These vertical gardens improve air quality, reduce building energy consumption, and provide a unique solution for space-constrained cities.

In summary, space optimization is a critical consideration in vertical farming. By utilizing innovative growing structures, modular systems, and space-efficient crop selections, vertical farms can maximize their productivity within limited urban spaces. Integrating vertical farming into existing urban infrastructure further enhances its potential to contribute to sustainable urban food systems.

Infrastructure and Logistics

The infrastructure and logistics of vertical farming are vital to its success, efficiency, and sustainability. These elements ensure that plants receive the necessary resources for growth, energy is utilized efficiently, and waste is managed responsibly. This section will cover the key aspects of water and nutrient delivery systems, energy sources and sustainability, and waste management, providing a comprehensive understanding of the infrastructure and logistical considerations in vertical farming.

Water and Nutrient Delivery Systems

Effective water and nutrient delivery systems are crucial for the health and productivity of crops in vertical farming. These systems must ensure that plants receive the right amount of water and nutrients while minimizing waste and optimizing resource use:

- Hydroponic Systems:
    - Nutrient Film Technique (NFT): In NFT systems, a thin film of nutrient solution flows continuously over the roots of plants. The roots absorb the necessary nutrients while exposed to air, providing oxygen. This highly efficient method conserves water, as the nutrient solution is recirculated within the system.
    - Deep Water Culture (DWC): In DWC systems, plants are suspended in net pots with their roots submerged in a nutrient-rich, oxygenated water reservoir. Air stones or diffusers oxygenate the water, ensuring roots receive adequate oxygen and nutrients. DWC is simple to set up and maintain, making it a popular choice for leafy greens and herbs.
    - Drip Systems: Drip systems deliver nutrient solution directly to the base of each plant through a network of

tubes and emitters. Excess solution is collected and recirculated, reducing waste. This method allows for precise control over nutrient delivery, ensuring each plant receives the appropriate nutrients.

- Aeroponic Systems:

Misting Systems: In aeroponic systems, plants are grown with their roots suspended in the air. Nutrient solution is delivered to the roots through misting devices, providing a fine mist that coats the roots. This method ensures roots receive a balanced supply of nutrients and oxygen, promoting rapid growth and high yields. Aeroponics uses less water than traditional hydroponics, making it an efficient option for vertical farming.

- Aquaponic Systems:

Symbiotic Relationships: Aquaponic systems combine hydroponics with aquaculture, creating a symbiotic relationship between plants and fish. Fish waste provides essential nutrients for the plants, while the plants help filter and purify the water for the fish. This closed-loop system recycles water and nutrients, reducing the need for chemical fertilizers and promoting sustainability. Aquaponics can produce crops and fish, offering a diverse range of products from a single system.

- Automation and Monitoring:

Automated Nutrient Delivery: Advanced vertical farms use automated systems to deliver plant nutrients and water. Sensors monitor the solution's nutrient levels, pH, and temperature, and automated dispensers adjust the concentration and distribution of nutrients based on real-time data. This ensures that plants receive consistent and optimal nutrition, reducing the risk of nutrient deficiencies or excesses.

- Water Recycling and Conservation:

  Vertical farms employ water recycling systems to conserve water and reduce waste. Nutrient solutions are collected, filtered, and recirculated within the system, minimizing water usage and ensuring efficient resource use. Advanced filtration technologies remove contaminants and pathogens, maintaining water quality and safety.

In conclusion, effective water and nutrient delivery systems are essential for vertical farming's success. By employing advanced hydroponic, aeroponic, and aquaponic methods and integrating automation and monitoring technologies, vertical farms can ensure that plants receive the necessary resources for optimal growth while conserving water and reducing waste.

Energy Sources and Sustainability

Energy consumption is a significant factor in the operation of vertical farms. To ensure sustainability and reduce environmental impact, vertical farms must utilize energy-efficient technologies and explore renewable energy sources:

- Energy-Efficient Technologies:
  - LED Lighting: LED (light-emitting diode) lighting is a cornerstone of energy efficiency in vertical farming. LEDs consume less energy than traditional lighting systems while providing the specific light spectrum for photosynthesis. They also generate less heat, reducing the need for additional cooling. Programmable LED systems allow for precise light intensity and duration control, optimizing energy use.
  - Climate Control Systems: Advanced climate control systems, including heating, ventilation, and air conditioning (HVAC), are designed to maintain stable growing conditions while minimizing energy consumption. These systems use energy-efficient

technologies such as heat exchangers, variable speed fans, and smart thermostats to regulate temperature, humidity, and airflow. Automated controls and sensors ensure that climate conditions are optimized for plant growth, reducing energy waste.

- Renewable Energy Sources:
  - o Solar Power: Solar panels can be installed on the rooftops or facades of vertical farms to harness solar energy. This renewable energy source can significantly reduce reliance on grid electricity and lower operating costs. Solar power is advantageous in regions with abundant sunlight, providing a sustainable and cost-effective energy solution.
  - o Wind Energy: In some locations, wind turbines can be integrated into vertical farm designs to generate renewable energy. Wind energy is a viable option for farms with consistent wind patterns. Small-scale turbines can be installed on rooftops or nearby land, contributing to the farm's energy needs.
  - o Energy Storage: Renewable energy systems can complement energy storage solutions like batteries. Energy storage ensures a stable and reliable power supply, even when renewable energy generation fluctuates due to weather conditions. Advanced battery technologies, such as lithium-ion batteries, store excess energy generated during peak production periods for use during low production periods.

- Sustainable Practices:
  - o Energy Audits and Optimization: Regular energy audits help identify areas for improvement in energy efficiency. Vertical farms can analyze energy usage patterns and implement optimization strategies to reduce consumption. This includes upgrading equipment, improving insulation, and implementing energy-saving practices.
  - o Carbon Footprint Reduction: Vertical farms can significantly reduce their carbon footprint using renewable energy sources and energy-efficient

31

technologies. These sustainable practices minimize environmental impact and promote a greener food production system.

- o Integration with Smart Grids: Vertical farms can integrate with smart grid systems to enhance energy management. Smart grids use advanced technologies to balance supply and demand, optimize energy distribution, and improve grid reliability. Vertical farms connected to smart grids can benefit from dynamic pricing, demand response programs, and improved energy efficiency.

In summary, energy efficiency and sustainability are critical considerations in operating vertical farms. Vertical farms can reduce their environmental impact by leveraging energy-efficient technologies, renewable energy sources, and sustainable practices and contribute to a more sustainable and resilient food production system.

Waste Management

Effective waste management is essential for the sustainability and environmental impact of vertical farming. Proper waste handling and recycling practices ensure that vertical farms minimize waste generation, reduce resource consumption, and promote environmental stewardship:

- Organic Waste Management:
  - o Composting: Organic waste, such as plant trimmings, leaves, and roots, can be composted to create nutrient-rich compost. This compost can enrich growing media or as a soil amendment in other agricultural applications. Composting reduces the volume of organic waste sent to landfills and recycles valuable nutrients into the farming system.
  - o Vermicomposting: Vermicomposting involves using earthworms to decompose organic waste into high-quality vermicompost. This method accelerates the

composting process and produces a nutrient-dense soil conditioner. Vermicomposting is particularly effective for managing small-scale organic waste and can be integrated into vertical farming operations.

- Nutrient Solution Management:
  - Recycling Nutrient Solutions: Nutrient solutions used in hydroponic, aeroponic, and aquaponic systems can be recycled and reused. Advanced filtration and purification systems remove contaminants and pathogens, allowing the nutrient solution to be recirculated within the system. This practice conserves water and nutrients, reducing waste and operational costs.
  - Safe Disposal: When nutrient solutions can no longer be recycled, safe disposal practices must be followed to prevent environmental contamination. This includes neutralizing the solution to safe pH levels and following local regulations to dispose of agricultural waste.
- Packaging and Plastic Waste:

Sustainable Packaging: Vertical farms can reduce plastic waste by using sustainable packaging materials for their produce. Biodegradable, compostable, and recyclable packaging options minimize environmental impact and appeal to environmentally conscious consumers. Reducing single-use plastics and reusable containers also contribute to sustainable waste management practices.

- Plastic Recycling:

Plastic components used in vertical farming systems, such as trays, pipes, and containers, can be recycled at the end of their life cycle. Implementing a plastic recycling program ensures that these materials are properly processed and repurposed, reducing the volume of plastic waste.

- Waste-to-Energy Solutions:

- Anaerobic Digestion: Organic waste can be converted into biogas through anaerobic digestion. This process involves microorganisms breaking down organic matter without oxygen, producing biogas (a renewable energy source) and digestate (a nutrient-rich byproduct). Biogas can generate electricity or heat, while digestate can be used as a fertilizer.
- Pyrolysis and Gasification: These advanced waste-to-energy technologies convert organic waste into energy through thermal processes. Pyrolysis involves heating organic waste without oxygen to produce biochar, bio-oil, and syngas. Gasification converts organic waste into syngas, which can be used to generate electricity or as a chemical feedstock. These technologies offer sustainable waste management solutions and contribute to energy production.

- E-Waste Management:

Recycling Electronic Components: Vertical farming systems rely on electronic devices, sensors, and automation equipment. Proper e-waste management ensures that electronic components are recycled and disposed of responsibly. Partnering with certified e-waste recycling facilities ensures that valuable materials are recovered and hazardous substances are managed safely.

In conclusion, effective waste management is a critical aspect of vertical farming sustainability. By implementing composting, nutrient solution recycling, sustainable packaging, waste-to-energy solutions, and responsible e-waste management, vertical farms can minimize waste generation, conserve resources, and promote environmental stewardship. These practices contribute to the overall sustainability and success of vertical farming in urban environments.

Chapter 3: Economic Aspects of Vertical Farming

The economic aspects of vertical farming play a crucial role in determining its feasibility, scalability, and long-term sustainability. Understanding the financial dynamics, market opportunities, and economic challenges associated with vertical farming is essential for investors, entrepreneurs, and policymakers. This chapter explores the key economic considerations of vertical farming, including cost analysis, market opportunities, and financial viability. By delving into these aspects, we can understand the economic landscape of vertical farming and its potential to transform urban agriculture and food production.

Cost Analysis

Understanding the costs associated with vertical farming is crucial for assessing its financial viability and planning successful operations. This section will provide a detailed analysis of the initial investment, operational costs, and overall financial viability of vertical farming. By breaking down these components, we can gain insights into the economic feasibility of establishing and maintaining a vertical farm.

Initial Investment

The initial investment required to set up a vertical farm is significant, encompassing various aspects such as infrastructure, technology, and system installation. The following points outline the major components of the initial investment:

- Infrastructure and Construction:
  - Building or Retrofitting: Purchasing or leasing a building or retrofitting an existing structure is one of the largest initial expenses. New constructions designed for vertical farming can be more expensive

35

but offer customized solutions. Retrofitting existing structures, such as warehouses or abandoned buildings, can be more cost-effective but may require significant modifications.

- o Structural Modifications: Retrofitting often involves reinforcing floors to support the weight of growing systems, installing new HVAC systems, and modifying electrical and plumbing systems to meet the demands of vertical farming.
- Technology and Equipment:
  - o Growing Systems: Investing in hydroponic, aeroponic, or aquaponic systems is a significant part of the initial investment. The cost varies based on the scale, type, and complexity of the system. High-quality materials and advanced technologies can raise costs but improve efficiency and productivity.
  - o Lighting Systems: LED lighting systems are essential for vertical farming, providing the necessary light spectrum for photosynthesis. The initial cost of LED lights can be high, but their energy efficiency and long lifespan make them a worthwhile investment.
  - o Climate Control Systems: Advanced HVAC systems, including heating, cooling, ventilation, and humidity control, are necessary to maintain optimal growing conditions. These systems involve significant upfront costs but are crucial for ensuring consistent crop quality and yield.
- Automation and Monitoring:
  - o Sensors and Control Systems: Installing IoT sensors and automated control systems for monitoring environmental conditions, nutrient delivery, and plant health is another significant expense. These technologies enhance efficiency and reduce labor costs but require a substantial initial investment.
  - o Software and Data Analytics: Purchasing software platforms for data collection, analysis, and decision-making support is also part of the initial investment. These platforms help optimize operations and improve resource management.

- Permits and Licenses:

  Regulatory Compliance: Obtaining the necessary permits and licenses for operating a vertical farm involves legal and administrative costs. Compliance with local regulations and building codes ensures smooth operations and avoids potential legal issues.

In conclusion, the initial investment for setting up a vertical farm is substantial, encompassing infrastructure, technology, and regulatory compliance costs. While the upfront expenses are high, these investments lay the foundation for efficient and sustainable operations, ultimately contributing to the long-term success of the vertical farm.

Operational Costs

Once a vertical farm is established, ongoing operational costs must be managed to ensure profitability and sustainability. These costs include energy, labor, maintenance, and other recurring expenses. The following points detail the major components of operational costs:

- Energy Costs:
  - Lighting: LED lighting systems, while energy-efficient, still contribute to a significant portion of the operational costs. The electricity required to power the lights, especially in large-scale operations, can be substantial. Implementing energy-saving practices and utilizing renewable energy sources can help mitigate these costs.
  - Climate Control: Operating HVAC systems to maintain optimal temperature, humidity, and ventilation also incurs significant energy costs. Efficient climate control systems and smart energy management practices are essential to minimize these expenses.

- Labor Costs:
  - Staffing: Labor costs include salaries for farm managers, technicians, and workers involved in planting, maintenance, and harvesting. Skilled labor is required to operate and maintain advanced vertical farming systems, and competitive wages are necessary to attract and retain qualified staff.
  - Training and Development: Investing in ongoing training and development programs for staff ensures they know about the latest technologies and best practices in vertical farming. While this adds to operational costs, it enhances efficiency and productivity.
- Maintenance and Repairs:
  - Equipment Maintenance: Regular maintenance of growing systems, lighting, and climate control equipment is essential to ensure smooth operations and prevent breakdowns. This includes cleaning, calibration, and replacement of parts.
  - System Upgrades: Periodic upgrades to technology and equipment are necessary to keep the farm competitive and efficient. These upgrades involve additional costs but can lead to long-term savings and improved performance.

- Consumables and Inputs:
  - Nutrients and Growing Media: The cost of nutrient solutions, growing media (such as rock wool or coco coir), and other inputs required for plant growth must be factored into operational expenses. Efficient management and recycling of these inputs can help reduce costs.
  - Water: Although vertical farming uses less water than traditional agriculture, water consumption still incurs costs. Efficient water management practices, including recycling and purification, help minimize water expenses.
- Miscellaneous Costs:

- Packaging and Distribution: Costs associated with packaging materials, transportation, and distribution of produce to markets or consumers are part of the operational expenses. Sustainable packaging options and efficient distribution networks can help manage these costs.
- Insurance and Security: Insurance coverage for the facility, equipment, and crops is necessary to protect against potential risks. Security measures to prevent theft or vandalism also contribute to operational expenses.

In summary, operational costs in vertical farming include energy, labor, maintenance, consumables, and miscellaneous expenses. Effective management of these costs is crucial for the financial sustainability of the vertical farm. Implementing energy-efficient technologies, training staff, and optimizing resource use can help reduce operational expenses and improve profitability.

Financial Viability

Assessing the financial viability of a vertical farm involves analyzing the potential return on investment (ROI), profitability, and economic sustainability. Understanding the factors that influence financial viability helps in making informed decisions and planning for long-term success:

- Revenue Streams:
  - Crop Sales: The primary revenue stream for vertical farms is the sale of fresh produce. High-value crops, such as leafy greens, herbs, and microgreens, typically generate significant revenue due to their market demand and premium prices. Diversifying the crop portfolio to include fruits, vegetables, and specialty crops can enhance revenue potential.
  - Value-Added Products: Vertical farms can increase profitability by producing value-added products, such as packaged salads, herb blends, or plant-based foods.

These products often command higher prices and appeal to health-conscious consumers.
- o Partnerships and Contracts: Establishing partnerships with local restaurants, grocery stores, and food service providers can provide stable revenue through contract agreements. These partnerships ensure consistent demand for produce and reduce market volatility.
- Cost Management:
  - o Efficiency Improvements: Implementing energy-efficient technologies, optimizing resource use, and automating processes contribute to cost savings. Continuous monitoring and analysis of operations help identify areas for improvement and cost reduction.
  - o Economies of Scale: Scaling up operations can lead to economies of scale, where the cost per production unit decreases as the farm size increases. Larger vertical farms can negotiate better input prices, reduce overhead costs, and achieve higher productivity.
- Market Factors:
  - o Consumer Demand: Understanding market demand and consumer preferences is crucial for financial viability. Vertical farms must stay attuned to trends in healthy eating, sustainability, and local sourcing to cater to market needs. Marketing strategies that highlight the benefits of fresh, locally grown produce can enhance consumer appeal.
  - o Competition: Analyzing the competitive landscape helps vertical farms position themselves effectively in the market. Differentiating products through quality, variety, and sustainability practices can provide a competitive edge.

- Risk Management:
  - o Diversification: Diversifying crops and revenue streams reduce dependence on a single product and mitigates risks associated with market fluctuations or crop failures. Vertical farms can explore additional

revenue sources, such as agritourism, workshops, and educational programs.
- o Insurance and Contingency Planning: Obtaining insurance coverage for crops, equipment, and facilities protects against unforeseen events, such as natural disasters or equipment breakdowns. Developing contingency plans ensures preparedness for potential disruptions.
- Funding and Investment:
  - o Access to Capital: Securing funding through loans, grants, or investment is essential for establishing and expanding vertical farming operations. Investors are increasingly interested in sustainable and innovative agricultural ventures, providing funding opportunities.
  - o Return on Investment (ROI): Calculating the ROI helps determine the financial feasibility of vertical farming projects. A positive ROI indicates that the revenues generated outweigh the costs, making the venture profitable. Long-term financial planning and forecasting are necessary to ensure sustained profitability.

In conclusion, the financial viability of vertical farming depends on effective revenue generation, cost management, market understanding, risk mitigation, and access to capital. By analyzing these factors and implementing strategic practices, vertical farms can achieve economic sustainability and contribute to the future of urban agriculture.

Market Opportunities

Vertical farming presents numerous market opportunities that can drive its growth and success. Understanding consumer demand, identifying niche markets, and leveraging competitive advantages are essential for capitalizing on these opportunities. This section will explore the various aspects of market opportunities, including consumer demand, niche markets, and competitive advantage, providing a comprehensive analysis of how vertical farms can thrive in the evolving agricultural landscape.

Consumer Demand

Consumer demand plays a critical role in the success of vertical farming. Understanding the factors that drive consumer preferences and behavior can help vertical farms tailor their products and marketing strategies to meet market needs. The following points outline key aspects of consumer demand in the context of vertical farming:

- Health and Wellness Trends:
  - Fresh and Nutritious Produce: Consumers increasingly prioritise fresh, nutritious, high-quality produce. Vertical farms can meet this demand by providing locally grown, pesticide-free, and nutrient-dense fruits and vegetables. The ability to harvest produce at peak ripeness and deliver it quickly to local markets ensures superior freshness and flavor.
  - Organic and Sustainable Options: There is a growing consumer interest in organic and sustainably produced food. Vertical farms can leverage their controlled environment agriculture (CEA) systems to produce crops without synthetic pesticides and with minimal environmental impact. Marketing produce as organic and sustainable can attract health-conscious and environmentally aware consumers.
- Convenience and Availability:
  - Year-Round Production: Vertical farming's ability to produce crops year-round, regardless of seasonal variations, ensures a consistent fresh produce supply. Consumers value the availability of their favorite fruits and vegetables throughout the year, and vertical farms can capitalize on this demand by offering a steady stream of products.
  - Urban Proximity: The location of vertical farms within urban areas provides a significant advantage in

terms of convenience. Consumers increasingly seek locally sourced food to reduce their carbon footprint and support local economies. Vertical farms can highlight their proximity to urban consumers, emphasizing the benefits of reduced transportation distances and fresher produce.

- Transparency and Traceability:
    - Food Safety and Quality: Consumers are more concerned than ever about food safety and the origins of their food. Vertical farms can offer transparency and traceability, providing detailed information about growing practices, nutrient management, and harvest dates. This transparency builds consumer trust and confidence in the safety and quality of the produce.
    - Storytelling and Branding: Sharing the story behind the farm and its sustainable practices can enhance consumer connection and loyalty. Vertical farms can use branding and storytelling to communicate their commitment to innovation, sustainability, and community engagement, creating a strong brand identity that resonates with consumers.

In conclusion, understanding consumer demand is essential for the success of vertical farming. By aligning their products with health and wellness trends, offering convenience and availability, and providing transparency and traceability, vertical farms can effectively meet market needs and drive consumer loyalty.

Niche Markets

Vertical farming offers unique opportunities to cater to niche markets often underserved by traditional agriculture. By identifying and targeting these niche markets, vertical farms can diversify their product offerings, enhance profitability, and create a strong market presence. The following points outline key niche markets for vertical farming:

Gourmet and Specialty Foods

- High-End Restaurants and Chefs: Gourmet restaurants and chefs seek high-quality, unique, and specialty ingredients to create innovative dishes. Vertical farms can supply exotic herbs, microgreens, edible flowers, and specialty vegetables not commonly available in traditional markets. Establishing partnerships with high-end restaurants and chefs can provide a stable and lucrative revenue stream.
- Artisanal Food Producers: Artisanal food producers, such as craft brewers, distillers, and specialty food makers, require unique and high-quality product ingredients. Vertical farms can grow specialty crops like hops, botanicals, and heirloom varieties to cater to these producers, adding value to their offerings.

Health and Wellness Market*

- Nutraceuticals and Functional Foods: The demand for nutraceuticals and functional foods—foods that offer health benefits beyond basic nutrition—is on the rise. Vertical farms can produce crops with enhanced nutritional profiles, such as high-antioxidant berries, omega-3 rich greens, and medicinal herbs. These products can be marketed to health-conscious consumers and companies in the health and wellness industry.
- Dietary Supplements: Vertical farms can also grow plants used in dietary supplements, such as spirulina, chlorella, and other superfoods. The controlled environment of vertical farming ensures consistent quality and potency, making it an attractive option for supplement manufacturers.

Ethnic and Cultural Foods

- Ethnic Grocers and Markets: Many urban areas have diverse populations with specific dietary preferences and cultural food traditions. Vertical farms can grow crops that cater to these communities, such as specialty vegetables, herbs, and fruits used in ethnic cuisines. Supplying ethnic grocers and markets with culturally significant produce can create a loyal customer base.

- Community Supported Agriculture (CSA) Programs: Vertical farms can establish CSA programs that offer culturally diverse produce boxes. By providing a variety of crops that reflect the culinary traditions of different communities, vertical farms can build strong relationships with local consumers and promote food diversity.

Education and Research

- Educational Institutions: Schools, universities, and research institutions want to incorporate vertical farming into their educational programs and research initiatives. Vertical farms can partner with these institutions to provide hands-on learning experiences, research sustainable agriculture, and develop new farming technologies.
- Public and Private Sector Collaboration: Government agencies and private companies increasingly invest in sustainable agriculture and food security. Vertical farms can collaborate with these entities on projects that address urban food production, climate resilience, and environmental sustainability, creating opportunities for funding and innovation.

In summary, targeting niche markets allows vertical farms to diversify their revenue streams and create a strong market presence. By catering to gourmet and specialty foods, health and wellness products, ethnic and cultural foods, and educational and research initiatives, vertical farms can enhance their profitability and market relevance.

Competitive Advantage

Vertical farming offers several competitive advantages that can set it apart from traditional agriculture and other urban farming methods. Leveraging these advantages can help vertical farms establish a strong market position, attract investment, and achieve long-term success. The following points outline the key competitive advantages of vertical farming:

Sustainability and Environmental Impact:

- Resource Efficiency: Vertical farming uses less water and land than traditional agriculture. Advanced hydroponic, aeroponic, and aquaponic systems recycle water and nutrients, reducing waste and conserving resources. Highlighting these sustainability benefits can appeal to environmentally conscious consumers and investors.
- Reduced Carbon Footprint: By growing food close to urban centers, vertical farms reduce the need for long-distance transportation, lowering greenhouse gas emissions. This local production model contributes to a smaller carbon footprint and aligns with growing consumer demand for sustainable and locally sourced food.

Consistent Quality and Yield:

- Controlled Environment: Vertical farms operate in controlled environments where temperature, humidity, light, and nutrient levels are precisely managed. This control ensures consistent crop quality and high yields, regardless of external weather conditions. The ability to produce reliable and high-quality produce year-round is a significant competitive advantage.
- Pest and Disease Management: The controlled environment of vertical farming minimizes the risk of pests and diseases, reducing the need for chemical pesticides and herbicides. This results in healthier and safer produce, which can be marketed as pesticide-free and environmentally friendly.

Technological Innovation:

- Advanced Farming Techniques: Vertical farms leverage cutting-edge technologies, including IoT, AI, and automation, to optimize operations and improve efficiency. These innovations enhance productivity, reduce labor costs, and ensure precise resource management. Positioning vertical farming as a high-tech solution can attract tech-savvy consumers and investors.
- Data-Driven Decision Making: The use of data analytics and real-time monitoring allows vertical farms to make informed decisions about crop management, resource allocation, and market strategies. This data-driven approach enhances operational efficiency and competitiveness.

Urban Integration and Proximity:

- Local Production: Vertical farms located in urban areas have the advantage of proximity to consumers. This local production model reduces transportation costs and time, ensuring fresher produce and faster delivery. Emphasizing the benefits of local sourcing can enhance consumer trust and loyalty.
- Urban Revitalization: Vertical farms can contribute to urban revitalization by repurposing unused spaces, such as rooftops, warehouses, and vacant buildings. This integration into the urban fabric not only provides fresh produce but also creates green spaces, jobs, and community engagement opportunities.

Market Differentiation:

- Unique Product Offerings: Vertical farms can differentiate themselves by offering unique and specialty crops that are not widely available in traditional markets. This includes exotic herbs, microgreens, edible flowers, and high-nutrient varieties. Differentiating products based on quality, sustainability, and innovation can create a competitive edge.
- Branding and Marketing: Effective branding and marketing strategies that highlight the benefits of vertical farming, such

as sustainability, freshness, and innovation, can attract consumers and build a loyal customer base. Storytelling and transparent communication about farming practices can enhance brand identity and market appeal.

In conclusion, vertical farming's competitive advantages lie in its sustainability, consistent quality, technological innovation, urban integration, and market differentiation. By leveraging these advantages, vertical farms can establish a strong market position, attract investment, and achieve long-term success in the evolving agricultural landscape.

Chapter 4: Environmental Impact

The environmental impact of vertical farming is a key consideration in evaluating its sustainability and potential to revolutionize agriculture. By leveraging advanced technologies and innovative farming practices, vertical farming offers a range of environmental benefits, including reduced water usage, decreased reliance on chemical inputs, and lower greenhouse gas emissions. This chapter will delve into the various environmental impacts of vertical farming, examining how it conserves resources, enhances biodiversity, and contributes to mitigating urban heat islands. Understanding these impacts provides valuable insights into the role of vertical farming in promoting sustainable urban development and addressing global environmental challenges.

Sustainability and Resource Efficiency

Sustainability and resource efficiency are foundational principles of vertical farming, aimed at minimizing environmental impact while maximizing productivity. Vertical farming systems are designed to optimize water, energy, and other resources, making them a more sustainable alternative to traditional agriculture. This section will explore the key aspects of sustainability and resource efficiency in vertical farming, focusing on water conservation, energy efficiency, and reducing carbon footprint.

Water Conservation

Water conservation is one of the most significant advantages of vertical farming over traditional agricultural methods. Conventional farming is known for its high water consumption, with irrigation accounting for approximately 70% of global freshwater withdrawals. Vertical farming, on the other hand, employs advanced techniques that dramatically reduce water usage.

Hydroponics and Aeroponics:

- Hydroponics: In hydroponic systems, plants are grown in a nutrient-rich water solution rather than soil. This method allows for precise control over water delivery, ensuring that plants receive the exact amount of water they need. The closed-loop nature of hydroponic systems means that water is recirculated and reused, significantly reducing overall water consumption. Studies have shown that hydroponic systems can use up to 90% less water than traditional soil-based agriculture.
- Aeroponics: Aeroponic systems take water conservation further by delivering nutrients to plants through a fine mist. In these systems, plant roots are suspended in the air and periodically misted with a nutrient solution. This method uses even less water than hydroponics, as the misting process ensures that only minimal amounts of water are needed to sustain plant growth. Aeroponic systems can reduce water usage by up to 98% compared to traditional farming.

Water Recycling and Management:

- Recirculation Systems: Vertical farms have recirculation systems that capture and reuse water. These systems include advanced filtration and purification technologies that remove impurities and pathogens from the water, making it safe for reuse. By continuously recycling water, vertical farms minimize waste and enhance water efficiency.
- Real-Time Monitoring: The use of IoT sensors and automation in vertical farming allows for real-time monitoring of water levels, quality, and usage. These sensors provide valuable data that can be used to optimize water management practices, ensuring that plants receive the right amount of water at the right time. Automated systems can adjust water delivery based on plant needs, reducing overwatering and water waste.

Climate Control Benefits:

- Reduced Evaporation: In controlled environment agriculture (CEA) systems, such as those used in vertical farming, the climate is carefully managed to reduce water loss through evaporation. The enclosed nature of vertical farms means that water remains within the system, further conserving this valuable resource. Humidity levels can be adjusted to minimize evaporation, ensuring more water is available for plant uptake.
- Efficient Plant Growth: The controlled environment of vertical farming promotes optimal plant growth, reducing the water required to produce the same amount of biomass as traditional farming. Faster growth rates and higher yields mean water is used more efficiently, contributing to overall sustainability.

In conclusion, vertical farming offers significant water conservation benefits through hydroponic and aeroponic systems, water recycling and management practices, and climate control technologies. These methods drastically reduce water usage compared to traditional farming, making vertical farming a sustainable solution for addressing global water scarcity and promoting resource efficiency.

Energy Efficiency

Energy efficiency is another critical aspect of sustainability in vertical farming. While vertical farms rely on energy-intensive technologies, such as artificial lighting and climate control systems, they also incorporate various strategies to minimize energy consumption and enhance overall efficiency.

LED Lighting:

- Energy-Efficient LEDs: Light-emitting diode (LED) lighting is a cornerstone of energy efficiency in vertical farming. LEDs consume significantly less energy than traditional incandescent or fluorescent lights while providing the specific light spectrum needed for photosynthesis. Their high energy efficiency, long lifespan, and low heat output make LEDs an ideal choice for indoor farming.

- Customized Light Spectrums: LEDs can be tailored to emit specific wavelengths of light that promote plant growth and development. This customization ensures that plants receive the optimal light conditions for photosynthesis, reducing the need for excessive lighting and energy consumption. Programmable LED systems can also simulate natural daylight cycles, enhancing energy efficiency.

Climate Control Systems:

- Efficient HVAC Systems: Heating, ventilation, and air conditioning (HVAC) systems are essential for maintaining optimal growing conditions in vertical farms. Advanced HVAC systems use energy-efficient technologies, such as heat exchangers, variable speed fans, and smart thermostats, to regulate temperature and humidity. These systems are designed to minimize energy use while ensuring that plants thrive in a stable environment.

- Automated Climate Control: Automation is crucial in optimizing energy use in vertical farms. Automated climate control systems, equipped with sensors and data analytics, continuously monitor environmental conditions and make real-time adjustments to heating, cooling, and ventilation. This precise control minimizes energy waste and enhances overall efficiency.

Renewable Energy Integration:

- Solar Power: Integrating solar panels into the design of vertical farms can significantly reduce reliance on grid electricity. Solar power provides a sustainable and renewable energy source that can offset the energy demands of lighting and climate control systems. Vertical farms in regions with abundant sunlight can benefit greatly from solar energy, reducing their carbon footprint and operational costs.

- Energy Storage Solutions: Renewable energy sources, such as solar and wind, can be complemented with energy storage systems, such as batteries, to ensure a stable and reliable power supply. Energy storage allows vertical farms to store excess energy generated during peak production periods and use it during periods of low production or high demand. This approach enhances energy security and efficiency.

Energy Audits and Optimization:

- Regular Energy Audits: Conducting regular energy audits helps vertical farms identify areas for improvement in energy efficiency. These audits analyze energy usage patterns, equipment performance, and operational practices to pinpoint inefficiencies and opportunities for savings. Implementing recommendations from energy audits can lead to significant reductions in energy consumption.
- Optimization Strategies: Vertical farms can implement various strategies to optimize energy use, such as improving insulation, using energy-efficient equipment, and implementing demand response programs. These strategies not only reduce energy consumption but also lower operational costs and enhance sustainability.

In summary, energy efficiency is a critical component of sustainability in vertical farming. By leveraging energy-efficient technologies, integrating renewable energy sources, and continuously optimizing energy use, vertical farms can minimize their environmental impact and enhance their overall sustainability. These practices contribute to the long-term viability of vertical farming as a sustainable agricultural solution.

Reduction of Carbon Footprint

Reducing the carbon footprint is a major goal of vertical farming, contributing to its role as a sustainable and environmentally friendly agricultural practice. Vertical farms employ various strategies to minimize greenhouse gas emissions and promote carbon neutrality.

Local Production and Reduced Transportation

- Urban Proximity: Vertical farms are often located in urban areas, close to the consumers they serve. This proximity significantly reduces the need for long-distance transportation, which is a major source of carbon emissions in traditional agriculture. By producing food near urban centers, vertical farms can shorten supply chains, reduce transportation costs, and lower the associated carbon footprint.
- Freshness and Shelf Life: Locally grown produce from vertical farms reaches consumers more quickly than food transported from distant rural farms. This reduced transportation time not only lowers emissions but also enhances the freshness and shelf life of the produce, reducing food waste and further contributing to sustainability.

Energy-Efficient Practices:

- Renewable Energy Use: As mentioned earlier, integrating renewable energy sources, such as solar and wind power, into vertical farming operations can significantly reduce reliance on fossil fuels. By harnessing clean energy, vertical farms can lower their carbon emissions and contribute to a more sustainable energy system.
- Optimized Energy Consumption: Using energy-efficient technologies and practices in vertical farms helps minimize energy consumption and associated emissions. LED lighting, automated climate control, and efficient HVAC systems reduce the energy required for crop production, contributing to a smaller carbon footprint.

Sustainable Agricultural Practices:

- Reduced Chemical Use: Vertical farming's controlled environment minimizes the need for chemical pesticides and herbicides, which are often derived from fossil fuels and contribute to greenhouse gas emissions. By reducing

chemical inputs, vertical farms lower their environmental impact and promote healthier ecosystems.
- Efficient Resource Use: Vertical farming's efficient use of water, nutrients, and growing media reduces waste and the environmental impact of resource extraction and processing. This efficiency contributes to a more sustainable agricultural system with a lower carbon footprint.

Carbon Sequestration and Offsetting:

- Carbon Sequestration: Vertical farms can contribute to carbon sequestration by cultivating certain plants that capture and store carbon dioxide. For example, some crops can produce biochar, a stable form of carbon that can be added to soil to improve fertility and sequester carbon over the long term.
- Carbon Offsetting Initiatives: Vertical farms can participate in carbon offsetting programs to compensate for emissions. By investing in projects that reduce or capture carbon emissions, such as reforestation or renewable energy initiatives, vertical farms can achieve carbon neutrality and enhance their sustainability credentials.

Lifecycle Analysis and Sustainability Reporting:

- Lifecycle Analysis (LCA): Conducting lifecycle analyses helps vertical farms assess the total environmental impact of their operations, from production to distribution. LCA provides insights into areas where emissions can be reduced and helps farms develop strategies for improving sustainability.
- Sustainability Reporting: Transparent reporting on sustainability practices and achievements helps vertical farms build trust with consumers, investors, and stakeholders.

By sharing data on carbon footprint reduction, resource efficiency, and environmental impact, vertical farms can demonstrate their commitment to sustainability and attract support for their initiatives.

In conclusion, reducing the carbon footprint is a key aspect of sustainability in vertical farming. By producing food locally, utilizing renewable energy, implementing energy-efficient practices, and participating in carbon offsetting initiatives, vertical farms can minimize their greenhouse gas emissions and contribute to a more sustainable agricultural system. These efforts position vertical farming as a vital component of the global strategy to address climate change and promote environmental stewardship.

Biodiversity and Ecosystems

Vertical farming is not only a solution to urban food production but also a means to enhance urban biodiversity and support ecosystems. By integrating nature into cityscapes, vertical farms can provide habitats for various species, deliver valuable ecosystem services, and help mitigate the urban heat island effect. This section will explore the impact of vertical farming on urban biodiversity, the ecosystem services it supports, and its role in reducing urban heat islands.

Urban Biodiversity

Urban biodiversity refers to the variety of living organisms in urban areas, including plants, animals, and microorganisms. Traditionally, urban environments are not conducive to high levels of biodiversity due to habitat fragmentation, pollution, and limited green spaces. However, vertical farming offers a unique opportunity to enhance urban biodiversity by creating new habitats and supporting diverse species.

Vertical farms can serve as mini-ecosystems within cities, providing habitats for various plant species and, indirectly, for insects, birds, and other wildlife. Including various plant species in vertical farms can promote pollinator diversity. Pollinators such as bees, butterflies, and other insects are attracted to flowering plants grown in vertical farms, which can provide them with food and shelter. This is particularly important in urban areas where pollinator habitats are often scarce.

Furthermore, vertical farms can be designed to include native plant species, which are well-adapted to local conditions and support local wildlife. By incorporating native plants, vertical farms contribute to regional biodiversity conservation and help maintain ecological balance. These native species can provide food and nesting sites for urban wildlife, enhancing the overall biodiversity of the area.

Vertical farms can also serve as green corridors that connect fragmented habitats in urban areas. These green corridors facilitate the movement of species between different habitats, promoting genetic diversity and resilience. For example, green walls and rooftop gardens can link parks, community gardens, and other green spaces, creating continuous habitats for wildlife.

In addition to supporting wildlife, vertical farms contribute to the microbial diversity of urban environments. The diverse plant species and soil substrates used in vertical farms harbor various microorganisms, including beneficial bacteria and fungi. These microorganisms are crucial in nutrient cycling, soil health, and plant growth. By promoting microbial diversity, vertical farms help create healthier urban ecosystems.

In conclusion, vertical farming enhances urban biodiversity by providing habitats for various plant and animal species, supporting pollinators, incorporating native plants, and creating green corridors. These efforts contribute to the overall health and resilience of urban ecosystems, making cities more livable and sustainable.

Ecosystem Services

Ecosystem services are the benefits that humans derive from natural ecosystems. These services include provisioning (such as food and water), regulating (such as climate regulation and air quality improvement), supporting (such as nutrient cycling and soil formation), and cultural (such as recreational and aesthetic) services. Vertical farming can provide several valuable ecosystem services in urban areas, enhancing the quality of life for residents and contributing to environmental sustainability.

One of the primary ecosystem services provided by vertical farming is food production. Vertical farms can produce various crops, including fruits, vegetables, herbs, and flowers, contributing to local food security and reducing the need for food transportation. By producing food locally, vertical farms help reduce the carbon footprint associated with food distribution and provide fresh, nutritious produce to urban residents.

Vertical farming also contributes to air quality improvement, a critical regulating service. Plants grown in vertical farms absorb carbon dioxide and release oxygen through photosynthesis, helping to reduce greenhouse gas concentrations in the atmosphere. Additionally, plants can trap dust, pollutants, and particulate matter, improving the air quality in urban areas. This is particularly beneficial in cities with high levels of air pollution, where vertical farms can act as green lungs, purifying the air and enhancing public health.

Another important ecosystem service provided by vertical farming is climate regulation. Plants help regulate urban temperatures through the processes of transpiration and shading. Transpiration involves the release of water vapor from plant leaves, which cools the air and mitigates the urban heat island effect. The shading provided by green walls and rooftop gardens also reduces the absorption of heat by buildings, lowering indoor temperatures and reducing the need for air conditioning.

Vertical farms support biodiversity by providing habitats for various species, as discussed earlier. This biodiversity contributes to ecosystem stability and resilience, enhancing the ability of urban ecosystems to withstand environmental stressors such as climate change, pollution, and habitat fragmentation. The presence of diverse plant species in vertical farms also supports pollinators and other beneficial insects, promoting ecological balance and contributing to crop pollination.

In addition to these regulating and supporting services, vertical farming offers cultural ecosystem services. Vertical farms can enhance the aesthetic appeal of urban environments, providing green spaces that improve the well-being of residents. Green walls, rooftop gardens, and urban farms create visually appealing landscapes, offering opportunities for recreation, relaxation, and community engagement. These green spaces can also serve as educational tools, raising awareness about sustainable agriculture and environmental stewardship.

In summary, vertical farming provides various ecosystem services, including food production, air quality improvement, climate regulation, biodiversity support, and cultural benefits. These services enhance the quality of life for urban residents and contribute to the sustainability and resilience of urban ecosystems.

Mitigating Urban Heat Island Effect

The urban heat island (UHI) effect refers to the phenomenon where urban areas experience higher temperatures than their rural surroundings. This effect is primarily caused by the extensive use of impervious surfaces such as asphalt, concrete, and buildings, which absorb and retain heat. The lack of vegetation in urban areas exacerbates the UHI effect, leading to increased energy consumption, higher pollution levels, and adverse health impacts. Vertical farming can significantly mitigate the UHI effect by introducing vegetation into urban environments and altering the microclimate.

One of the key ways vertical farming mitigates the UHI effect is through transpiration. Transpiration is the release of water vapor from plant leaves into the atmosphere. This process cools the air and increases humidity, creating a microclimate that counteracts the heat retained by urban structures. Vertical farms, with their dense plantings, can significantly increase transpiration in urban areas, reducing ambient temperatures and alleviating the UHI effect.

In addition to transpiration, vertical farms provide shading, which helps lower temperatures in urban environments. Green walls and rooftop gardens absorb less heat than traditional building materials, preventing heat from being trapped and radiated back into the urban environment. The shading effect of vertical farms reduces the temperature of building surfaces and the surrounding air, contributing to cooler urban microclimates.

Furthermore, vertical farms can insulate buildings, reducing the need for air conditioning and lowering energy consumption. The vegetation on green walls and rooftops acts as a natural insulator, keeping buildings cooler in the summer and warmer in the winter. This insulation effect decreases the reliance on mechanical cooling and heating systems, saving energy and reducing greenhouse gas emissions. By lowering energy demand, vertical farms indirectly contribute to the mitigation of the UHI effect.

Integrating vertical farms into urban planning and architecture also promotes the creation of green corridors, which enhance air circulation and reduce heat buildup. Green corridors are interconnected green spaces that facilitate air movement and help dissipate heat. By incorporating vertical farms into these green corridors, cities can improve ventilation, enhance airflow, and reduce the overall temperature of urban areas.

Moreover, vertical farms can contribute to the overall greening of cities, increasing the amount of green space per capita. This greening effect not only mitigates the UHI effect but also provides additional benefits such as improved air quality, enhanced biodiversity, and increased recreational opportunities. Green spaces have been shown to improve mental health and well-being, providing urban residents a connection to nature and a respite from the built environment.

In conclusion, vertical farming is crucial in mitigating the urban heat island effect by promoting transpiration, providing shading, insulating buildings, enhancing air circulation, and increasing urban green spaces. These efforts help lower urban temperatures, reduce energy consumption, and improve the overall livability of cities. By integrating vertical farms into urban planning and development, cities can create cooler, more sustainable, and resilient environments that support the well-being of their residents.

Chapter 5: Social and Cultural Implications

The social and cultural implications of vertical farming extend far beyond its technological and environmental benefits. As vertical farming becomes more integrated into urban landscapes, it influences community dynamics, lifestyle choices, and cultural practices. This chapter explores how vertical farming fosters community engagement, enhances food security, and shapes urban cultural landscapes. By examining these social and cultural dimensions, we can better understand the broader impacts of vertical farming on urban life and society, highlighting its potential to create more connected, resilient, and vibrant communities.

Community Engagement and Education

Vertical farming has the potential to transform urban communities by fostering engagement and providing educational opportunities. By integrating vertical farms into the fabric of cities, we can create spaces that bring people together, promote learning, and enhance the overall quality of urban life. This section explores how vertical farming contributes to community engagement through urban agriculture programs, educational initiatives, and community gardens.

Urban Agriculture Programs

Urban agriculture programs are an effective way to engage communities in vertical farming. These programs provide opportunities for residents to participate in food production, learn about sustainable practices, and connect with their neighbors. Vertical farming can play a central role in these programs, offering a unique and innovative approach to urban agriculture:

- Participation and Inclusion: Urban agriculture programs that incorporate vertical farming can engage a diverse range of participants, including students, seniors, and individuals from various socioeconomic backgrounds. By providing accessible and inclusive opportunities for involvement, these programs can foster a sense of community and shared purpose. Participants can participate in planting, maintaining, and harvesting crops, gaining hands-on experience and a deeper understanding of food production.

- Workshops and Training: Vertical farming-based urban agriculture programs can offer workshops and training sessions on various topics, such as hydroponics, aeroponics, and sustainable agriculture. These educational opportunities enable participants to acquire new skills and knowledge, empowering them to adopt sustainable practices in their own lives. Workshops can also cover composting, water conservation, and organic pest management, further promoting environmental stewardship.
- Health and Nutrition: Urban agriculture programs that include vertical farming can improve community health and nutrition by providing access to fresh, locally grown produce. Participants can learn about the nutritional benefits of different crops and how to incorporate them into their diets. Additionally, these programs can host cooking classes and demonstrations, teaching residents how to prepare healthy, delicious meals using the produce they have grown.
- Economic Opportunities: Vertical farming-based urban agriculture programs can create economic opportunities for community members. By establishing local markets or cooperatives, participants can sell their produce, generating income and supporting the local economy. These programs can also offer job training and employment opportunities in urban agriculture, helping to address issues of unemployment and underemployment in urban areas.

In conclusion, urban agriculture programs incorporating vertical farming can engage communities meaningfully, providing opportunities for participation, education, improved health, and economic development. By fostering a sense of community and shared responsibility, these programs contribute to the overall well-being and resilience of urban neighborhoods.

Educational Initiatives

Vertical farming offers many educational opportunities that can benefit students, educators, and the broader community. By integrating vertical farms into educational settings, we can create engaging, hands-on learning experiences that promote sustainability and environmental awareness. This section explores the various educational initiatives that can be supported by vertical farming:

- School Programs: Vertical farming can be incorporated into school curriculums at all levels, from elementary to high school. By establishing vertical farms on school grounds, students can engage in experiential learning, exploring topics such as plant biology, ecology, and sustainable agriculture. Vertical farms can serve as living laboratories where students conduct experiments, observe plant growth, and learn about the science and technology behind modern farming methods.
- STEM Education: Vertical farming provides a unique platform for promoting STEM (science, technology, engineering, and mathematics) education. Students can learn about the principles of hydroponics and aeroponics, the role of sensors and automation in agriculture, and the impact of environmental factors on plant growth. These hands-on experiences can spark interest in STEM subjects and inspire future careers in science and technology.
- Higher Education and Research: Colleges and universities can integrate vertical farming into their programs and research initiatives. Vertical farms can support research on sustainable agriculture, climate change, and urban food systems. Students and faculty can collaborate on projects that explore innovative farming techniques, crop optimization, and resource management. Additionally, vertical farming can be incorporated into courses in environmental science, biology, engineering, and urban planning.
- Public Workshops and Events: Educational initiatives can extend beyond formal education settings to include public workshops and events. Vertical farms can host open houses, tours, and demonstrations, providing opportunities for

community members to learn about sustainable agriculture and food production. Workshops on home hydroponics, organic gardening, and composting can empower individuals to adopt sustainable practices in their own lives.

- Partnerships with Educational Institutions: Vertical farms can form partnerships with educational institutions to develop joint programs and initiatives. These partnerships can include collaborative research projects, internships, and service-learning opportunities. By working together, vertical farms and educational institutions can maximize their impact and reach a broader audience.

In summary, educational initiatives incorporating vertical farming can provide valuable learning experiences for students, educators, and the community. By promoting sustainability and environmental awareness, these initiatives contribute to the development of a more informed and engaged society. Vertical farming can serve as a powerful tool for education, inspiring future generations to embrace sustainable practices and contribute to a more resilient world.

Community Gardens

Community gardens are a vital component of urban agriculture, providing spaces where residents can come together to grow food, share knowledge, and build community. Vertical farming can enhance the impact of community gardens by introducing innovative growing techniques and increasing the diversity of crops that can be cultivated. This section explores the role of vertical farming in community gardens and its benefits for urban communities:

- Maximizing Space: One of the primary advantages of vertical farming in community gardens is the efficient use of space. In densely populated urban areas, space for traditional gardening is often limited. Vertical farming allows for the cultivation of crops in vertically stacked layers, significantly increasing the growing area within a small footprint. This space efficiency enables more residents to participate in community gardening and access fresh produce.

- Year-Round Growing: Vertical farming can extend the growing season for community gardens, allowing for year-round production of fruits and vegetables. By using controlled environment agriculture (CEA) techniques, such as hydroponics and aeroponics, community gardens can maintain optimal growing conditions regardless of external weather. This ensures a consistent supply of fresh produce, enhancing food security and providing residents with nutritious options throughout the year.
- Diverse Crop Selection: Vertical farming expands the range of crops grown in community gardens. Traditional community gardens may be limited to certain crops due to soil quality, space constraints, or climate conditions. Vertical farms can cultivate various plants, including leafy greens, herbs, microgreens, and even fruits and vegetables that may not thrive in traditional soil-based gardens. This diversity allows community members to explore new foods and incorporate a broader range of nutrients into their diets.
- Engaging Activities: Vertical farming introduces new and engaging activities for community garden participants. Residents can learn about advanced growing techniques, such as hydroponics and aeroponics, and gain hands-on experience with cutting-edge farming technologies. These activities foster a sense of curiosity and innovation, encouraging participants to experiment with different crops and growing methods.
- Social Interaction and Collaboration: Community gardens incorporating vertical farming provide opportunities for interaction and collaboration. Residents can work together to plan, plant, and maintain the vertical farm, fostering a sense of camaraderie and shared responsibility. These interactions help build stronger community bonds and create a supportive network of individuals who share a common interest in sustainable living.

- Health and Well-Being: Participating in community gardening has been shown to have numerous health benefits, including increased physical activity, improved mental health, and reduced stress levels. Vertical farming enhances these benefits by providing a productive and stimulating environment for gardening. The physical activity in vertical farming, such as planting, harvesting, and maintaining the systems, contributes to overall health and well-being.

In conclusion, vertical farming can significantly enhance the impact of community gardens by maximizing space, enabling year-round growing, diversifying crop selection, and providing engaging activities. These benefits contribute to stronger community bonds, improved health, and greater food security for urban residents. By integrating vertical farming into community gardens, cities can create vibrant, sustainable spaces that support the well-being of their residents and foster a sense of community.

Cultural Shifts and Acceptance

The integration of vertical farming into urban environments is not just a technological and environmental innovation but also a catalyst for cultural change. As vertical farming becomes more prevalent, it influences how people perceive food, shapes urban lifestyles, and drives policy and advocacy efforts. This section explores the cultural shifts and acceptance of vertical farming, focusing on changing food perceptions, urban lifestyle integration, and policy and advocacy.

Changing Food Perceptions

Vertical farming transforms how people perceive food, from its origins to its nutritional value and environmental impact. As consumers become more aware of the benefits of vertical farming, their attitudes toward food production and consumption are evolving.

Local and Sustainable Food

One of the significant cultural shifts driven by vertical farming is the increased appreciation for locally sourced and sustainably produced food. Consumers are becoming more conscious of the environmental impact of food transportation and the importance of supporting local agriculture. Vertical farms, often located within or near urban centers, provide fresh produce with a lower carbon footprint. This local production model resonates with consumers who prioritize sustainability and seek to reduce their environmental impact.

Freshness and Nutritional Value

Vertical farming emphasizes the production of fresh, nutrient-dense produce. Because vertical farms can deliver produce to local markets quickly after harvest, the food retains more nutritional value than items that travel long distances. Consumers recognize the health benefits of eating fresher produce, leading to a shift in preferences toward foods grown in vertical farms. This awareness is fostering a culture of health and wellness, where people prioritize the quality and nutritional content of their food.

Transparency and Trust

Vertical farming offers greater transparency in food production, which is increasingly important to consumers. Many vertical farms provide detailed information about their growing practices, including the absence of pesticides and sustainable methods. This transparency builds consumer trust and confidence in the safety and quality of the food. As a result, there is a growing preference for produce from vertical farms over conventionally grown food, which may lack such transparency.

Innovation and Modern Agriculture

The innovative nature of vertical farming is changing how people view agriculture. Traditionally, farming has been associated with rural areas and labor-intensive practices. Vertical farming, using advanced technologies such as hydroponics, aeroponics, and IoT, presents a modern, high-tech image of agriculture. This shift particularly appeals to younger generations, more inclined to embrace technological advancements and innovative solutions. As vertical farming gains acceptance, it reshapes the cultural narrative around agriculture, making it more appealing and relevant to urban populations.

In conclusion, vertical farming drives cultural shifts in food perceptions by promoting local and sustainable food, emphasizing freshness and nutritional value, offering transparency and building trust, and presenting a modern image of agriculture. These changes foster a culture that values quality, sustainability, and innovation in food production.

Urban Lifestyle Integration

As vertical farming becomes more integrated into urban settings, it influences urban lifestyles and transforms how people interact with their environment. The integration of vertical farming into cities is creating new opportunities for urban living, enhancing the quality of life, and promoting sustainable practices.

Accessibility and Convenience

Vertical farms, often located in or near urban centers, make fresh produce more accessible to city dwellers. This proximity reduces the need for long trips to farmers' markets or rural areas to obtain fresh food. The convenience of having vertical farms within urban neighborhoods encourages residents to incorporate more fresh produce into their diets. Urbanites can easily access fresh vegetables and herbs, either through local markets supplied by vertical farms or through direct sales from the farms themselves.

Green Spaces and Well-Being

Vertical farms contribute to creating green spaces in urban areas, enhancing the aesthetic appeal and livability of cities. Green walls, rooftop gardens, and urban farms provide residents access to nature, which has been shown to improve mental health and well-being.

These green spaces offer opportunities for recreation, relaxation, and social interaction, fostering a sense of community and connection to nature. Vertical farms in cities can transform urban environments, making them more pleasant and conducive to healthy living.

Community Involvement and Engagement

The integration of vertical farming into urban lifestyles promotes community involvement and engagement. Many vertical farms offer opportunities for residents to participate in farming activities, such as planting, harvesting, and maintaining the farms. Community-supported agriculture (CSA) programs, farm-to-table initiatives, and urban farming workshops encourage residents to participate actively in food production. This involvement fosters a sense of ownership and responsibility for local food systems, strengthening community bonds and promoting civic engagement.

Sustainable Living Practices

Vertical farming encourages urban residents to adopt more sustainable living practices. By providing education on sustainable agriculture, water conservation, and waste reduction, vertical farms can inspire individuals to make environmentally conscious choices in their daily lives. The visibility of vertical farms in urban areas is a constant reminder of the importance of sustainability, influencing behaviors such as recycling, composting, and reducing food waste. As vertical farming becomes more integrated into urban lifestyles, it promotes a culture of sustainability and environmental stewardship.

In summary, the integration of vertical farming into urban lifestyles enhances accessibility to fresh produce, creates green spaces, promotes community involvement, and encourages sustainable living practices. These changes contribute to a higher quality of life for urban residents and support the development of resilient, sustainable cities.

Policy and Advocacy

The growth and acceptance of vertical farming are closely tied to supportive policies and advocacy efforts. As vertical farming continues to gain traction, it is essential to develop policies that facilitate its expansion and promote its benefits. Advocacy efforts are crucial in raising awareness, influencing public opinion, and driving policy changes that support vertical farming.

Regulatory Frameworks

Developing a supportive regulatory framework is essential for the growth of vertical farming. This includes zoning laws, building codes, and agricultural regulations that accommodate the unique needs of vertical farms. Policymakers must recognize vertical farming as a legitimate form of agriculture and create policies that facilitate its establishment and operation within urban areas. This may involve updating existing regulations or introducing new ones that address the specific challenges and opportunities of vertical farming.

Incentives and Funding

Providing incentives and funding for vertical farming initiatives can encourage investment and innovation in the sector. Government grants, tax credits, and low-interest loans can help offset the initial costs of setting up vertical farms and support ongoing operations. Public-private partnerships can also be vital in funding research and development projects, promoting technological advancements, and scaling up successful vertical farming models. These incentives can drive growth in the vertical farming industry and ensure its long-term sustainability.

Education and Outreach

Advocacy efforts must focus on educating the public, policymakers, and stakeholders about the benefits of vertical farming. This includes raising awareness about its environmental, economic, and social advantages. Public outreach campaigns can highlight success stories, demonstrate the impact of vertical farming on local communities, and showcase the innovative technologies used in these farms. Engaging with media, organizing events, and collaborating with educational institutions can amplify these efforts and build widespread support for vertical farming.

Community Engagement and Participation

Effective advocacy involves engaging communities and encouraging their participation in vertical farming initiatives. Community involvement can be fostered through urban agriculture programs, workshops, and volunteer opportunities. By involving residents in the planning and development of vertical farms, advocacy efforts can ensure that these projects meet the needs and preferences of the local population. Community support is crucial for the success and sustainability of vertical farming, making it essential to involve residents in decision-making processes.

Collaboration with Stakeholders

Advocacy for vertical farming requires collaboration with various stakeholders, including government agencies, non-profit organizations, industry leaders, and academic institutions. Building alliances and partnerships can strengthen advocacy efforts and create a unified voice supporting vertical farming. Stakeholders can work together to address challenges, share best practices, and develop strategies for scaling up vertical farming initiatives. Collaborative efforts can also lead to the creation of comprehensive policies and programs that support the growth of vertical farming.

In conclusion, policy and advocacy are critical components of the cultural acceptance and expansion of vertical farming. Developing supportive regulatory frameworks, providing incentives and funding, educating the public, engaging communities, and collaborating with stakeholders are essential for promoting the benefits of vertical farming and ensuring its long-term success. By driving policy changes and fostering widespread support, advocacy efforts can help integrate vertical farming into the cultural and social fabric of urban environments.

Chapter 6: Challenges and Solutions

While vertical farming offers numerous benefits and opportunities, it also faces several challenges that must be addressed to ensure its success and sustainability. These challenges range from technological and financial hurdles to social and regulatory obstacles. This chapter explores the key challenges facing vertical farming and presents practical solutions. By understanding and addressing these challenges, vertical farming can continue to grow and thrive, contributing to sustainable urban agriculture and food security.

Technical Challenges

Vertical farming, while promising, is not without its technical challenges. Overcoming these challenges is crucial to ensuring the efficiency, productivity, and sustainability of vertical farms. This section explores the key technical challenges related to infrastructure and maintenance, pest and disease management, scalability issues, and potential solutions.

Infrastructure and Maintenance

One of the primary technical challenges in vertical farming is related to the infrastructure and maintenance of the farming systems. Building and maintaining the physical structures required for vertical farming can be complex and costly.

Structural Integrity

Ensuring the structural integrity of vertical farming facilities is essential. The buildings must support the weight of the growing systems, water, and plants. This often requires significant investment in reinforced structures and regular inspections to prevent potential failures.

Investing in robust construction materials and engineering expertise can ensure the facilities are built to last. Regular maintenance checks and structural assessments can identify and address potential issues before they become serious problems.

Climate Control Systems

Vertical farms rely heavily on climate control systems to maintain optimal growing conditions. These systems include heating, ventilation, air conditioning (HVAC), and humidity control. Any failure in these systems can lead to crop losses and reduced productivity.

Implementing advanced climate control technologies with automated monitoring and backup systems can minimize the risk of system failures. Regular maintenance and timely repairs of HVAC systems are crucial to ensure uninterrupted operation.

Lighting Systems

Artificial lighting, particularly LED lights, is a critical component of vertical farming. Ensuring consistent and efficient lighting is essential for plant growth. However, lighting systems can be expensive to install and maintain.

Using energy-efficient LED lights and optimizing their placement and usage can reduce operational costs. Regularly cleaning and maintaining lighting fixtures can extend their lifespan and ensure they provide adequate illumination.

Water and Nutrient Delivery Systems

Hydroponic and aeroponic systems require precise water and nutrient delivery. Clogging, leaks, or malfunctions in these systems can disrupt plant nutrient supply, leading to poor growth or crop failure.

Regular inspection and maintenance of water and nutrient delivery systems are necessary to prevent and address issues promptly. Using high-quality components and implementing automated monitoring systems can enhance the reliability of these systems.

In conclusion, addressing the infrastructure and maintenance challenges in vertical farming requires significant investment in robust construction, advanced climate control, efficient lighting, and reliable water and nutrient delivery systems. Regular maintenance and proactive management are essential to ensure the smooth operation and long-term success of vertical farms.

Pest and Disease Management

Pest and disease management is another critical challenge in vertical farming. Despite the controlled environment, vertical farms are not immune to pests and diseases, which can spread rapidly and cause significant damage.

Pest Infestation

Pests such as aphids, mites, and whiteflies can infiltrate vertical farms and damage crops. The enclosed environment of vertical farms can facilitate the rapid spread of pests if not managed effectively.

Implementing integrated pest management (IPM) strategies can effectively control pest populations. IPM involves a combination of biological controls (such as introducing beneficial insects), physical barriers, and minimal use of chemical pesticides. Regular monitoring and early detection are crucial to managing pest infestations promptly.

Disease Outbreaks

Fungal, bacterial, and viral diseases can also threaten vertical farms. High humidity levels, often maintained for optimal plant growth, can create favorable conditions for developing fungal diseases.

Maintaining optimal humidity levels and ensuring proper air circulation can help prevent fungal diseases. Using disease-resistant plant varieties and practicing good sanitation can reduce the risk of disease outbreaks. Implementing a comprehensive disease management plan that includes regular monitoring, early detection, and appropriate treatments is essential.

Sanitation and Hygiene

Ensuring proper sanitation and hygiene within the vertical farm is vital to prevent the introduction and spread of pests and diseases. Contaminated tools, equipment, and personnel can inadvertently introduce pathogens.

Establishing strict sanitation protocols, including regular cleaning and disinfection of tools and equipment, can minimize the risk of contamination. Implementing biosecurity measures, such as controlled access and requiring personnel to follow hygiene practices, can further protect the farm from pests and diseases.

Biological Control

Relying solely on chemical pesticides can lead to resistance and environmental concerns. Biological control methods, such as natural predators or biopesticides, offer sustainable alternatives.

Integrating biological control methods into pest and disease management plans can reduce the reliance on chemical pesticides and promote a healthier ecosystem within the vertical farm. Researching and adopting new biocontrol agents can enhance the effectiveness of these strategies.

In summary, effective pest and disease management in vertical farming requires a multifaceted approach that includes integrated pest management, disease prevention measures, strict sanitation practices, and biological control methods. These strategies help protect crops, maintain productivity, and promote sustainable farming practices.

Scalability Issues

Scalability is a significant challenge for vertical farming, as expanding operations can introduce new complexities and increase costs. Ensuring that vertical farms can scale effectively while maintaining efficiency and profitability is crucial for their long-term success.

Capital Investment

Scaling up vertical farming operations requires substantial capital investment in infrastructure, technology, and equipment. Securing funding for expansion can be challenging, especially for small-scale farms.

Developing a solid business plan and demonstrating the potential return on investment can attract investors and secure funding. Exploring public-private partnerships and grant opportunities can also provide financial support for scaling up operations.

Technological Integration

As vertical farms scale, integrating advanced technologies such as automation, IoT, and data analytics becomes increasingly important. These technologies can enhance efficiency, reduce labor costs, and improve crop management.

Investing in scalable technologies and modular systems that can be easily expanded is essential. Collaborating with technology providers and researchers can help identify and implement the best solutions for large-scale vertical farming operations.

Operational Complexity

Scaling up vertical farming operations increases the complexity of managing the farm. Larger operations require more sophisticated logistics, supply chain management, and workforce coordination.

Implementing robust management systems and utilizing data-driven decision-making can streamline operations and improve efficiency. Training and developing a skilled workforce capable of managing the complexities of large-scale vertical farming is also critical.

Market Demand

Ensuring that there is sufficient market demand to support scaled-up operations is essential. Expanding too quickly without a corresponding increase in market demand can lead to oversupply and financial losses.

Conducting thorough market research and developing a strong marketing strategy can help identify and tap into new markets. Building partnerships with retailers, restaurants, and other buyers can secure long-term contracts and ensure a stable demand for produce.

Resource Management

Scaling up vertical farming operations requires efficient ly managing resources such as water, energy, and nutrients. Larger farms must ensure that resource use remains sustainable and cost-effective.

Implementing advanced resource management systems and adopting sustainable practices can optimize resource use and reduce costs. Utilizing renewable energy sources and recycling water and nutrients can enhance the sustainability of large-scale vertical farming operations.

In conclusion, addressing scalability issues in vertical farming involves securing capital investment, integrating advanced technologies, managing operational complexity, ensuring market demand, and optimizing resource management. By tackling these challenges, vertical farms can scale effectively, maintain efficiency, and achieve long-term success in the agricultural industry.

Economic and Social Challenges

Vertical farming, while innovative and promising, faces significant economic and social challenges that can impact its widespread adoption and long-term viability. Addressing these challenges is essential for successfully integrating vertical farming into urban environments and broader agricultural systems. This section delves into the economic and social challenges of vertical farming, focusing on high initial costs, market competition, and social acceptance.

High Initial Costs

One of the most significant economic challenges facing vertical farming is the high initial cost required to establish and operate a vertical farm. These costs can be a barrier to entry for many potential farmers and investors.

Capital Investment

Setting up a vertical farm involves substantial capital investment in infrastructure, technology, and equipment. This includes the cost of building or retrofitting facilities, installing advanced climate control and lighting systems, and integrating automation and monitoring technologies.

To mitigate high initial costs, vertical farms can seek funding through various sources, including venture capital, government grants, and low-interest loans. Public-private partnerships can also provide financial support and share the risk of high initial investments. Developing a comprehensive business plan demonstrating the potential return on investment can attract investors and secure the necessary capital.

Operating Expenses

In addition to the initial setup costs, vertical farms face ongoing operating expenses such as energy, water, labor, and maintenance. These recurring costs can strain the financial sustainability of vertical farming operations.

Implementing energy-efficient technologies, such as LED lighting and advanced HVAC systems, can reduce operating expenses. Utilizing renewable energy sources like solar and wind power can further decrease energy costs. Efficient water management systems, including recycling and purification, can minimize water expenses. Automation and data analytics can optimize labor use and reduce the need for manual intervention, thereby lowering labor costs.

Economies of Scale

Achieving economies of scale can be challenging for vertical farms, especially smaller operations. Larger farms can benefit from reduced costs per production unit, while smaller farms may struggle to achieve similar efficiencies.

Scaling up operations can help vertical farms achieve economies of scale and reduce per-unit costs. Collaborating with other vertical farms to share resources, technology, and knowledge can also enhance efficiency and cost-effectiveness. Forming cooperatives or networks can enable smaller farms to leverage collective bargaining power and access bulk purchasing discounts.

In summary, addressing the high initial costs of vertical farming requires securing diverse funding sources, implementing cost-saving technologies, and achieving economies of scale. These strategies can help make vertical farming more financially accessible and sustainable.

Market Competition

Market competition is another significant economic challenge for vertical farming. Competing with traditional agriculture and other urban farming methods can be difficult, particularly in pricing and market share.

Competitive Pricing

Traditional agriculture often benefits from lower production costs due to economies of scale and established infrastructure. Vertical farms, with their higher initial and operating costs, may struggle to compete on price with conventionally grown produce.

Vertical farms can differentiate themselves by focusing on unique advantages, such as fresher, pesticide-free produce, and year-round availability. Highlighting the environmental and health benefits of their products can justify premium pricing. Additionally, optimizing production processes to reduce costs and improve efficiency can help vertical farms become more competitive in pricing.

Market Saturation

As vertical farming gains popularity, the risk of market saturation increases. A growing number of vertical farms can lead to increased competition and potential oversupply in certain markets.

Conducting thorough market research and identifying niche markets can help vertical farms avoid oversaturation. Diversifying crop varieties and offering unique, high-value products can attract specific customer segments and reduce direct competition. Building strong relationships with local retailers, restaurants, and consumers can create loyal customer bases and ensure steady demand.

Brand Recognition and Consumer Trust

Established agricultural brands have the advantage of consumer trust and recognition. New vertical farming operations may face challenges in building brand awareness and consumer trust.

Effective branding and marketing strategies are crucial for vertical farms to establish a strong market presence. Highlighting the quality, freshness, and sustainability of their products can attract environmentally conscious and health-focused consumers. Participating in community events, farmers' markets, and educational programs can increase visibility and build trust within the community.

In conclusion, market competition poses significant challenges for vertical farming. By differentiating their products, targeting niche markets, and building strong brands, vertical farms can overcome these challenges and successfully compete in the agricultural market.

Social Acceptance

Social acceptance is a crucial factor in the widespread adoption of vertical farming. Overcoming skepticism and gaining the support of the public, policymakers, and other stakeholders is essential for the success of vertical farming initiatives.

Public Perception

Many people are unfamiliar with vertical farming and may be skeptical about its benefits and feasibility. Misconceptions about the safety, quality, and sustainability of vertically farmed produce can hinder acceptance.

Public education and outreach are essential to address misconceptions and build awareness about the benefits of vertical farming. Hosting open houses, farm tours, and community workshops can provide firsthand experiences and demonstrate the farming methods used. Collaborating with local schools, universities, and media can amplify educational efforts and reach a broader audience.

Cultural Preferences

Traditional food cultures and preferences can influence acceptance of new agricultural methods. Some consumers may prefer conventionally grown produce due to familiarity or cultural significance.

Emphasizing the quality and freshness of vertically farmed produce can help overcome cultural resistance. Offering a variety of crops, including culturally significant ones, can cater to diverse consumer preferences. Engaging with community leaders and cultural organizations can build trust and support for vertical farming within different cultural groups.

Policy and Regulatory Support

Gaining the support of policymakers and regulatory bodies is critical for successfully implementing vertical farming projects. Regulatory hurdles, such as zoning laws and agricultural policies, can impede the development of vertical farms.

Advocacy and collaboration with policymakers are essential to create a supportive regulatory environment for vertical farming. Presenting evidence-based research on the benefits of vertical farming can influence policy decisions. Working with industry associations and advocacy groups can strengthen efforts to promote favorable policies and remove regulatory barriers.

Community Integration

Ensuring that vertical farming projects align with community needs and values is important for gaining local support. Projects that fail to consider community input may face resistance or lack of engagement.

Involving the community in the planning and developing vertical farming projects can enhance acceptance and support. Conducting surveys, holding public meetings, and creating advisory committees can gather valuable input and ensure that projects address local priorities. Transparent communication and regular updates on project progress can build trust and foster a sense of ownership among community members.

In summary, social acceptance is vital for the success of vertical farming. Through public education, cultural sensitivity, policy advocacy, and community integration, vertical farms can overcome social challenges and gain widespread support. These efforts can ensure that vertical farming becomes a valued and accepted part of urban agriculture and food production.

In conclusion, addressing the economic and social challenges of vertical farming involves mitigating high initial costs, navigating market competition, and fostering social acceptance. By implementing effective strategies in these areas, vertical farms can overcome barriers and achieve sustainable growth and success.

Chapter 7: Future of Vertical Farming

The future of vertical farming holds immense potential to revolutionize urban agriculture and address global food security challenges. As technology advances and sustainability becomes increasingly prioritized, vertical farming is poised to play a significant role in the future food systems. This chapter explores the anticipated developments in vertical farming, including technological advancements, scaling up and global impact, and the integration with smart cities and resilient urban planning. Examining these future trends, we can understand how vertical farming will evolve and contribute to a sustainable and food-secure world.

Technological Advancements

Technological advancements are at the heart of the evolution and future potential of vertical farming. Continuous innovation in agriculture technology will drive the efficiency, productivity, and sustainability of vertical farms. This section explores the anticipated future technologies, ongoing research and development efforts, and the integration of vertical farming with smart cities.

Future Technologies

The future of vertical farming is heavily influenced by the development and implementation of cutting-edge technologies. These technologies aim to enhance the precision, efficiency, and scalability of vertical farms.

Advanced Automation and Robotics

Automation and robotics will play a crucial role in the future of vertical farming. Automated systems can handle planting, monitoring, maintenance, and harvesting, significantly reducing labor costs and increasing efficiency. Robotics, such as autonomous drones and robotic arms, can perform delicate tasks precisely, ensuring consistent quality and yield.

Investing in advanced automation and robotics can streamline vertical farming operations, making them more efficient and cost-effective. As technology advances, these systems will become more accessible and affordable for vertical farms of all sizes.

Artificial Intelligence and Machine Learning

AI and machine learning will revolutionize data analysis and decision-making in vertical farming. AI algorithms can analyze vast amounts of data from sensors and monitoring systems to optimize growing conditions, predict crop yields, and identify potential issues before they become problems.

Integrating AI and machine learning into vertical farming systems can enhance precision agriculture, leading to better resource management and higher productivity. Continuous learning and adaptation will allow these systems to improve, further optimizing farm operations.

Innovative Lighting Solutions

Developing more efficient and adaptable lighting solutions will be pivotal for vertical farming. Future lighting technologies, such as quantum dot LEDs and tunable spectrum LEDs, will give plants the exact light spectrum they need for each growth stage, enhancing photosynthesis and growth rates.

Adopting these advanced lighting solutions can improve energy efficiency and crop yields. The ability to tailor light conditions to specific plant needs will result in healthier plants and higher-quality produce.

Enhanced Climate Control Systems

Future climate control systems will incorporate advanced sensors and smart technologies to maintain optimal growing conditions. These systems will be able to adjust temperature, humidity, and CO_2

levels in real-time, ensuring the best possible environment for plant growth.

Enhanced climate control systems will lead to more consistent and reliable crop production. These systems will reduce energy consumption and operational costs while maximizing plant health and productivity.

In conclusion, the future of vertical farming will be shaped by adopting advanced automation, AI, innovative lighting, and enhanced climate control systems. These technologies will drive efficiency, productivity, and sustainability, making vertical farming more viable and attractive for urban agriculture.

Research and Development

Ongoing research and development (R&D) efforts are essential for advancing vertical farming technologies and practices. By investing in R&D, vertical farms can innovate and improve their operations, contributing to the overall growth and success of the industry.

Crop Optimization

Research into optimizing crop varieties for vertical farming is critical. Developing plant varieties that thrive in controlled environments, have shorter growth cycles, and require fewer resources can significantly enhance productivity and sustainability.

Collaborating with agricultural research institutions and universities can drive the development of optimized crop varieties. Funding and supporting research initiatives focused on vertical farming can yield new insights and breakthroughs in crop science.

Nutrient Solutions and Growing Media

Research into nutrient solutions and growing media can lead to more efficient and sustainable farming practices. Developing nutrient solutions that maximize plant growth while minimizing waste and environmental impact is a key area of focus. Similarly, identifying or creating growing media that provide optimal plant support and nutrition is essential.

Conducting experiments and trials to test different nutrient solutions and growing media can help identify the most effective options. Sharing findings with the broader vertical farming community can accelerate the adoption of best practices.

Integrated Pest Management

Developing effective and sustainable pest management strategies is crucial for the success of vertical farming. Research into biological controls, natural predators, and other non-chemical pest management methods can reduce reliance on pesticides and promote healthier crops.

Investing in research to develop integrated pest management (IPM) strategies tailored to vertical farming environments can enhance crop protection and sustainability. Collaborating with entomologists and pest management experts can yield innovative solutions.

Sustainability and Resource Efficiency

Researching ways to improve the sustainability and resource efficiency of vertical farming is vital. This includes exploring renewable energy sources, water recycling techniques, and waste reduction strategies. The goal is to create closed-loop systems that minimize environmental impact.

Partnering with environmental scientists and sustainability experts can lead to the development of innovative solutions for resource efficiency. Implementing findings from sustainability research can reduce the environmental footprint of vertical farms.

In summary, ongoing research and development efforts are essential for advancing vertical farming technologies and practices. By focusing on crop optimization, nutrient solutions, pest management, and sustainability, vertical farms can innovate and improve their operations, contributing to the overall growth and success of the industry.

Integration with Smart Cities

Integrating vertical farming with smart cities represents a significant opportunity for enhancing urban sustainability and resilience. Smart cities leverage technology and data to improve urban living, and vertical farming can play a key role in this vision.

Data-Driven Urban Agriculture

Smart cities utilize data and analytics to optimize various aspects of urban living. Integrating vertical farming with smart city infrastructure allows for data-driven decision-making in urban agriculture. Real-time data on climate conditions, energy use, and crop performance can enhance the efficiency and productivity of vertical farms.

Implementing IoT sensors and data analytics platforms within vertical farms can provide valuable insights and improve farm management. Collaborating with smart city planners and technology providers can facilitate the integration of vertical farming into the broader smart city ecosystem.

Energy and Resource Management

Smart cities optimise energy and resource use to create sustainable urban environments. Vertical farms can contribute to this goal by utilizing renewable energy sources, recycling water, and reducing waste. Integrating vertical farming with smart grid systems and resource management platforms can enhance sustainability.

Investing in renewable energy solutions, such as solar panels and wind turbines, can reduce the energy footprint of vertical farms. Implementing water recycling and waste reduction techniques can improve resource efficiency. Collaborating with smart city initiatives focused on sustainability can amplify these efforts.

Urban Planning and Green Spaces

Vertical farming can contribute to creating green spaces and sustainable urban planning. Integrating vertical farms into buildings, rooftops, and public spaces can enhance the aesthetic appeal and livability of cities. These green spaces can provide recreational opportunities, improve air quality, and reduce the urban heat island effect.

Working with urban planners and architects to design vertical farms that complement urban landscapes can create multifunctional green spaces. Participating in smart city projects focused on green infrastructure can promote the integration of vertical farming into urban planning.

Community Engagement and Education

Smart cities prioritize community engagement and education to build resilient and connected communities. Vertical farming can play a role in these efforts by providing educational opportunities and engaging residents in sustainable agriculture. Community-supported agriculture (CSA) programs and urban farming workshops can foster a sense of community and environmental stewardship.

Developing educational programs and community engagement initiatives incorporating vertical farming can enhance public awareness and support. Collaborating with smart city initiatives focused on community engagement can amplify the impact of these efforts.

In conclusion, integrating vertical farming with smart cities presents significant opportunities for enhancing urban sustainability and resilience. By leveraging data-driven decision-making, optimizing resource use, contributing to green spaces, and engaging communities, vertical farming can play a vital role in the future of urban living. This integration will not only improve the efficiency and productivity of vertical farms but also contribute to the overall sustainability and livability of smart cities.

Scaling Up and Global Impact

The potential for vertical farming to scale up and impact global food systems is immense. As the world faces increasing challenges related to food security, urbanization, and climate change, scaling up vertical farming operations can provide sustainable solutions. This section explores expansion strategies, the impact on global food security, and the potential challenges that need to be addressed to realize the global potential of vertical farming.

Expansion Strategies

Scaling up vertical farming involves expanding existing operations and establishing new farms in various locations. Effective expansion strategies are crucial for ensuring that vertical farming can meet growing food demands and operate efficiently on a larger scale.

Franchise and Partnership Models

One effective strategy for scaling up vertical farming is to adopt franchise and partnership models. These models allow for rapid expansion by leveraging local expertise and resources while maintaining consistent standards and practices.

Developing a comprehensive franchise model can help standardize vertical farming operations and ensure quality control. Establishing partnerships with local entrepreneurs, businesses, and organizations can facilitate the expansion process and reduce the risks associated with entering new markets.

Modular and Scalable Systems

Designing vertical farming systems that are modular and easily scalable can simplify the expansion process. Modular systems can be easily replicated and adapted to different locations, allowing for flexible and efficient growth.

Investing in modular vertical farming systems that can be scaled up or down based on demand can enhance operational efficiency. These systems should be designed to integrate seamlessly with existing infrastructure and allow for easy expansion without significant disruptions.

Urban Integration

Integrating vertical farms into urban infrastructure, such as repurposed buildings, rooftops, and unused spaces, can facilitate expansion in densely populated areas. Urban integration not only provides fresh produce to local communities but also enhances the sustainability and livability of cities.

Collaborating with urban planners, architects, and real estate developers can identify suitable locations for vertical farms and streamline the integration process. Utilizing urban spaces effectively can maximize the impact of vertical farming on local food systems.

Technological Innovation

Leveraging technological advancements can support the scaling up of vertical farming operations. Innovations in automation, AI, and data analytics can enhance efficiency, reduce labor costs, and improve overall productivity.

Investing in cutting-edge technologies and continuously improving farming practices can enable vertical farms to scale efficiently. Collaborating with technology providers and research institutions can drive innovation and ensure vertical farms remain competitive and sustainable.

Market Diversification

Diversifying the range of products and markets served by vertical farms can reduce risks and enhance resilience. Expanding into new product categories, such as specialty crops and value-added products, can create additional revenue streams and meet diverse consumer needs.

Conducting market research to identify emerging trends and consumer preferences can inform product diversification strategies. Building strong relationships with retailers, food service providers, and direct-to-consumer channels can expand market reach and support sustainable growth.

In conclusion, effective expansion strategies, including franchise models, modular systems, urban integration, technological innovation, and market diversification, are essential for scaling up vertical farming operations. These strategies can ensure that vertical farming meets growing food demands and operates efficiently on a larger scale.

Global Food Security

Vertical farming has the potential to significantly impact global food security by providing a sustainable and reliable source of fresh produce. As the world population continues to grow, and climate change threatens traditional agricultural systems, vertical farming can be crucial in ensuring food security.

Year-Round Production

Vertical farming allows for year-round production of crops, regardless of external weather conditions. This consistent supply of fresh produce can help stabilize food markets and reduce the impact of seasonal fluctuations on food availability.

Implementing controlled environment agriculture (CEA) technologies in vertical farms can maintain optimal growing conditions year-round. This consistency in production can enhance food security by providing fresh produce to urban and rural communities.

Urban Food Production

As urbanization increases, the demand for locally grown food within cities rises. Vertical farming can bring food production closer to urban consumers, reducing the need for long-distance transportation and ensuring fresher produce.

Establishing vertical farms in urban centers can reduce the carbon footprint associated with food transportation and improve access to fresh produce. Urban vertical farms can also alleviate pressure on rural agricultural lands, preserving them for other essential uses.

Reducing Food Waste

Vertical farming can help reduce food waste by shortening the supply chain and minimizing the time between harvest and consumption. Fresh produce from vertical farms can reach consumers more quickly, reducing spoilage and waste.

Developing efficient distribution networks that connect vertical farms directly with local markets, restaurants, and consumers can minimize food waste. Implementing advanced packaging and preservation technologies can further extend the shelf life of fresh produce.

Resilience to Climate Change

Traditional agriculture is vulnerable to the impacts of climate change, such as extreme weather events, droughts, and pests. Vertical farming, with its controlled environments, is more resilient to these disruptions, ensuring a stable food supply.

Investing in climate-resilient technologies and practices can enhance the ability of vertical farms to withstand environmental stressors. Researching and developing crop varieties that thrive in controlled environments can improve resilience.

Addressing Food Deserts

Food deserts are areas with limited access to affordable and nutritious food. Vertical farming can help address this issue by establishing farms in underserved communities, improving access to fresh produce.

Identifying and targeting food deserts for vertical farming projects can enhance food security. Collaborating with community organizations and local governments can support the development and sustainability of these initiatives.

In summary, vertical farming can significantly impact global food security by providing year-round production, enhancing urban food production, reducing food waste, increasing resilience to climate change, and addressing food deserts. These contributions can help ensure a stable and sustainable food supply for growing populations worldwide.

Potential Challenges

While vertical farming holds great promise, it also faces several potential challenges that must be addressed to realize its full global impact. These challenges include economic, technical, and social barriers that can hinder the expansion and effectiveness of vertical farming operations.

Economic Viability

The high initial costs and ongoing operating expenses of vertical farming can pose significant economic challenges. Ensuring the financial sustainability of vertical farms is crucial for their long-term success.

Developing scalable and cost-effective technologies can reduce the financial barriers to entry. Securing diverse funding sources, such as government grants, private investments, and public-private partnerships, can support the economic viability of vertical farms. Implementing efficient resource management practices can also help reduce operating expenses.

Technological Integration

Integrating advanced technologies, such as automation, AI, and IoT, into vertical farming systems can be complex and costly. Ensuring seamless integration and effective use of these technologies is essential for maximizing productivity and efficiency.

Investing in research and development to improve the integration and scalability of advanced technologies can enhance their effectiveness. Collaborating with technology providers and experts can ensure vertical farms stay at the forefront of innovation. Training and support for farm operators can also facilitate the adoption of new technologies.

Regulatory and Policy Barriers

Regulatory and policy barriers can impede the development and expansion of vertical farming. Navigating zoning laws, agricultural regulations, and food safety standards can be challenging for vertical farms.

Advocacy and collaboration with policymakers are essential to create a supportive regulatory environment for vertical farming.

Developing clear guidelines and standards for vertical farming operations can streamline regulatory processes. Engaging with industry associations and advocacy groups can amplify efforts to promote favorable policies.

Social Acceptance and Cultural Adaptation

Gaining social acceptance and adapting to cultural preferences are crucial for successfully adopting vertical farming. Resistance to change and skepticism about new farming methods can hinder acceptance.

Public education and outreach efforts can address misconceptions and build awareness about the benefits of vertical farming. Engaging with communities and incorporating their input into vertical farming projects can enhance acceptance. Culturally relevant crops and products can also help integrate vertical farming into diverse communities.

Environmental Sustainability

Ensuring the environmental sustainability of vertical farming is critical. While vertical farming can reduce land and water use, it also relies heavily on energy and technology.

Implementing renewable energy sources and energy-efficient technologies can reduce the environmental footprint of vertical farms. Developing closed-loop systems that recycle water and nutrients can enhance sustainability. Regular environmental impact assessments can help identify areas for improvement and ensure that vertical farming practices remain sustainable.

In conclusion, addressing the potential challenges of vertical farming, including economic viability, technological integration, regulatory barriers, social acceptance, and environmental sustainability, is essential for realizing its full global impact. By implementing effective solutions and strategies, vertical farming can overcome these challenges and contribute to a more sustainable and food-secure future.

Conclusion

Summary of Key Points

Vertical farming represents a transformative approach to urban agriculture, offering innovative solutions to the most pressing challenges facing global food systems today. Throughout this book, we have explored various aspects of vertical farming, from its technological foundations and economic aspects to its environmental impact and social implications. This summary highlights the key points discussed in each chapter, emphasizing the major themes and overall benefits of vertical farming.

Recap of Major Themes

In the initial chapters, we delved into the basics of vertical farming, defining the concept and exploring its historical background and current trends. Vertical farming involves growing crops in vertically stacked layers within controlled environments, utilizing advanced technologies such as hydroponics, aeroponics, and aquaponics. This method addresses the limitations of traditional agriculture by optimizing space, reducing water usage, and enhancing crop yields.

The importance of vertical farming was highlighted through its contributions to food security, efficient land use in urban areas, and environmental sustainability. Vertical farms can produce food year-round, regardless of external weather conditions, ensuring a consistent fresh produce supply. By utilizing urban spaces, vertical farming reduces the need for extensive agricultural land, preserving natural ecosystems and promoting biodiversity.

Technological advancements play a crucial role in the success of vertical farming. Automation, artificial intelligence (AI), and the Internet of Things (IoT) enable precise control of growing conditions, enhancing productivity and resource efficiency. Advanced lighting systems, climate control technologies, and nutrient delivery methods ensure optimal plant growth, while ongoing research and development drive continuous innovation.

Economic aspects were also thoroughly examined, focusing on cost analysis, market opportunities, and financial viability. The high initial costs of setting up vertical farms can be mitigated through various funding sources, including venture capital, government grants, and public-private partnerships. Market opportunities arise from the growing demand for fresh, locally sourced produce, as well as niche markets for specialty foods and functional crops. Achieving financial viability requires effective cost management, economies of scale, and strategic market positioning.

Environmental impact is a significant consideration in vertical farming. This approach offers numerous benefits, including water conservation, energy efficiency, and reduced carbon footprint. Vertical farms use significantly less water than traditional agriculture, thanks to recirculating hydroponic systems and advanced irrigation techniques. Energy-efficient LED lighting and renewable energy sources contribute to sustainability, while local production minimizes the environmental impact of food transportation.

Social and cultural implications were explored, emphasizing community engagement, education, and cultural shifts. Vertical farming fosters community involvement through urban agriculture programs, educational initiatives, and community gardens. It promotes social acceptance and integration by addressing public perceptions, cultural preferences, and policy support. As vertical farming becomes more integrated into urban lifestyles, it enhances the quality of life for residents and supports sustainable living practices.

Future prospects for vertical farming include technological advancements, scaling up operations, and global impact. Continued innovation in automation, AI, and climate control will drive the efficiency and productivity of vertical farms. Scaling up involves effective expansion strategies, modular systems, and urban integration. Vertical farming has the potential to address global food security by providing a reliable source of fresh produce, reducing food waste, and enhancing resilience to climate change.

Overall Benefits of Vertical Farming

Vertical farming offers numerous benefits that make it a compelling solution for sustainable urban agriculture. These benefits span environmental, economic, and social dimensions, contributing to a more resilient and food-secure future.

Environmentally, vertical farming significantly reduces the resource intensity of food production. By utilizing controlled environment agriculture, vertical farms can achieve higher yields with less water and land than traditional farming methods. Hydroponics and aeroponics minimizes water waste, while advanced climate control systems optimize energy use. Additionally, vertical farms can be powered by renewable energy sources, reducing their carbon footprint. Localized food production reduces the need for long-distance transportation, cutting greenhouse gas emissions and enhancing the freshness and quality of produce.

Economically, vertical farming creates new opportunities for investment, innovation, and job creation. The initial costs of establishing vertical farms can be offset by various funding mechanisms, while ongoing operational efficiencies drive profitability. Vertical farming supports local economies by generating employment and fostering entrepreneurship in urban agriculture. Market opportunities for fresh, high-quality produce, specialty crops, and value-added products further enhance economic viability. By diversifying revenue streams and targeting niche markets, vertical farms can build resilient business models that withstand market fluctuations.

Socially, vertical farming fosters community engagement, education, and well-being. Urban agriculture programs and community gardens provide spaces for residents to connect, learn, and participate in food production. Educational initiatives at schools and universities promote sustainability awareness and inspire future generations of urban farmers. Vertical farms enhance urban aesthetics by creating green spaces that improve mental health and provide recreational opportunities. These farms also address food deserts by improving access to nutritious, locally grown food in underserved communities.

Vertical farming promotes cultural shifts towards sustainable living and healthy eating. By emphasizing the benefits of fresh, pesticide-free produce, vertical farming encourages consumers to make healthier food choices. The transparency and traceability of vertical farming practices build consumer trust and confidence in food safety and quality. As vertical farming becomes more integrated into urban planning and development, it supports the creation of smart, sustainable cities that prioritize environmental stewardship and social equity.

In conclusion, vertical farming offers a holistic approach to addressing the challenges of modern agriculture. Its environmental benefits include water conservation, energy efficiency, and reduced carbon emissions. Economically, vertical farming drives innovation, supports local economies, and creates new market opportunities. Socially, it fosters community engagement, enhances education, and promotes healthy, sustainable living. By leveraging advanced technologies and sustainable practices, vertical farming has the potential to transform urban agriculture and contribute to a more resilient and food-secure world.

Call to Action

The potential of vertical farming to revolutionize urban agriculture and contribute to global food security is immense, but realizing this potential requires collective effort and proactive engagement. This section outlines a call to action, emphasizing the importance of engaging stakeholders and promoting sustainable practices to ensure the successful integration and expansion of vertical farming.

Engaging Stakeholders

The success of vertical farming depends on the active involvement and collaboration of a diverse range of stakeholders, including policymakers, investors, technology providers, educators, and community members. Each group plays a crucial role in driving the development and adoption of vertical farming.

Policymakers

To create a supportive environment for vertical farming, policymakers must develop and implement regulations that accommodate the unique needs of this innovative agricultural method. This includes updating zoning laws, building codes, and agricultural policies to facilitate the establishment and operation of vertical farms. Policymakers should also provide financial incentives, such as grants, tax credits, and low-interest loans, to support the initial investment and ongoing operations of vertical farms. By prioritizing vertical farming in urban planning and development strategies, policymakers can promote sustainable urban growth and food security.

Investors

Securing adequate funding is essential for the expansion of vertical farming. Investors, including venture capitalists, impact investors, and public funding bodies, must recognize the potential of vertical farming to address global food challenges and support its growth. Investing in vertical farming not only provides financial returns but also contributes to social and environmental benefits. Investors should focus on funding scalable and sustainable vertical farming projects, as well as supporting research and development efforts to drive innovation in the sector.

Technology Providers

Advanced technologies are the backbone of vertical farming. Technology providers, including companies specializing in automation, AI, climate control, and LED lighting, must continue to innovate and develop solutions tailored to the needs of vertical farms. Collaboration between technology providers and vertical farming operators is crucial to ensure that the latest advancements are effectively integrated into farming systems. Technology providers should also offer training and support to help farm operators maximize the benefits of new technologies.

Educators

Education is vital in raising awareness about vertical farming and building the skills needed to operate and manage vertical farms. Educational institutions at all levels, from primary schools to universities, should incorporate vertical farming into their curriculums and provide hands-on learning experiences. Research institutions should prioritize studies on vertical farming, exploring new techniques and solutions to improve efficiency and sustainability. By educating future generations and conducting groundbreaking research, educators can drive the growth and development of vertical farming.

Community Members

Community engagement is essential for successfully integrating vertical farming into urban environments. Local communities must be involved in the planning and developing of vertical farming projects to ensure that these initiatives meet their needs and preferences. Community members can participate in urban agriculture programs, volunteer at community gardens, and support local vertical farms by purchasing their produce. Building strong relationships between vertical farms and their communities can enhance social acceptance and support for this innovative agricultural method.

In conclusion, engaging stakeholders is crucial for the success of vertical farming. Policymakers, investors, technology providers, educators, and community members all have a role to play in driving the development and adoption of vertical farming. By working together, these stakeholders can create a supportive environment that promotes sustainable urban agriculture and contributes to global food security.

Promoting Sustainable Practices

To maximize the environmental and social benefits of vertical farming, it is essential to promote and adopt sustainable practices. Sustainable vertical farming practices can reduce resource consumption, minimize environmental impact, and enhance the resilience and long-term viability of vertical farms.

Resource Efficiency

Vertical farms must prioritize efficiently using resources, including water, energy, and nutrients. Implementing advanced water management systems, such as hydroponics and aeroponics, can significantly reduce water usage compared to traditional agriculture. Recycling and reusing water within the farm can further enhance water efficiency. Energy-efficient technologies, such as LED lighting and advanced climate control systems, can minimize energy consumption. Integrating renewable energy sources, such as solar and wind power, can reduce the carbon footprint of vertical farms. Nutrient management systems should be optimized to deliver precise amounts of nutrients to plants, minimizing waste and environmental impact.

Waste Reduction

Reducing waste is a key component of sustainable vertical farming. Vertical farms should implement strategies to minimize organic waste, such as composting plant residues and using them as soil amendments. Recycling and reusing growing media can also reduce waste. Vertical farms should adopt packaging solutions that are biodegradable, recyclable, or reusable to reduce plastic waste. Developing closed-loop systems that recycle nutrients and water can further enhance sustainability and reduce the environmental footprint of vertical farming.

Biodiversity and Ecosystem Services

Promoting biodiversity within vertical farms can enhance ecosystem services and resilience. Incorporating a diverse range of plant species can support pollinators and other beneficial insects, contributing to ecological balance. Vertical farms can also provide habitats for urban wildlife, promoting urban biodiversity. Implementing integrated pest management (IPM) strategies that use biological controls and minimize chemical pesticide use can enhance the health and sustainability of vertical farming systems.

Community and Economic Benefits

Sustainable vertical farming practices should also focus on maximizing community and economic benefits. Engaging local communities in vertical farming initiatives can create educational and employment opportunities, enhancing social equity and well-being. Supporting local economies by sourcing materials and services from local suppliers can strengthen community resilience. Developing value-added products, such as ready-to-eat salads and herbal teas, can create additional revenue streams and meet diverse consumer needs.

Continuous Improvement and Innovation

Sustainable vertical farming requires continuous improvement and innovation. Vertical farms should invest in research and development to explore new technologies, techniques, and practices that enhance sustainability. Collaborating with research institutions, technology providers, and other vertical farms can drive innovation and share best practices. Regular monitoring and assessment of environmental performance can help identify areas for improvement and ensure that vertical farming practices remain sustainable.

In summary, promoting sustainable practices is essential for maximizing the environmental and social benefits of vertical farming. By prioritizing resource efficiency, waste reduction, biodiversity, community and economic benefits, and continuous improvement, vertical farms can enhance their sustainability and contribute to a more resilient and food-secure world. Adopting these practices can ensure that vertical farming remains a viable and sustainable solution for urban agriculture in the future.

www.ingramcontent.com/pod-product-compliance
Lightning Source LLC
Chambersburg PA
CBHW070423240526
45472CB00020B/1170